WATER RESOURCES AND THEIR UTILISATION

S.C. Kalwar, Ex-Professor and Head, Department of Geography, Member, Board of Studies in Geography, Ex-Member of Academic Council and Senate, University of Rajasthan, Jaipur is a specialist in Agricultural Geography, Rural Development, Land use Planning, Economic Geography, Wasteland Development Studies, Environment and Surveying. He has written/edited 18 books related to resources, wasteland, ecology, environment, poverty and socio-economic development and has participated in about 36 conferences/seminars/symposia of National and International level. He has contributed 65 research papers in various journals of repute. He is a member of Editorial Board of 'Studies in Geography' and member of an Advisory Committee of 'Charanmvati', a research journal.

Prof. Kalwar was President of the All Rajasthan Geographers Association (RGA) in 2005-06 and is a life member of National Association of Geographers of India, Rajasthan Geographers Association, Asian Environment Council and Indian Institute of Geomorphology. He was also a Member of National Committee, Government of India for evaluating Hindi Books on environment for giving prize to the authors at National level, Ministry of Home Affairs, Government of India, New Delhi (2002). He was selected for a Research Award (Career Award) in 1992 by the University Grants Commission, New Delhi.

Prof. Kalwar is a co-ordination consultant in Geography, Vardhman Mahaveer Open University, Kota (Raj.). He has prepared 21 books for UG, PG and M.Phil. students of VMOU, Kota during last two years.

Amit Kanthoj did his B.Sc., M.A., Ph.D. and MCA from University of Rajasthan, Jaipur. Dr. Kanthoj is a lecturer and incharge of Computer Center, Agarwal Post Graduate College, Jaipur. He has written many books of Computer Science, for the degree and diploma courses in computer for various Universities.

S.C. Kalwar, Ex-Professor and Head, Department of Geography, Member, Board of Studies in Geography, Ex-Member of Academic Council and Senate, University of Rajasthan, Jaipur, is a specialist in Agricultural Geography, Rural Development, Land use Planning, Economic Geography, Wasteland Development Studies, Environment and Surveying. He has written/edited 18 books related to resources, wasteland, ecology, environment, poverty and socio-economic development and has participated in about 36 conferences/seminars/symposia of National and International level. He has contributed 66 research papers in various journals of repute. He is a member of Editorial Board of 'Studies in Geography' and member of an Advisory Committee of 'Charamavati', a research journal.

Prof. Kalwar was President of the All Rajasthan Geographers Association (RGA) in 2005-06 and is a life member of National Association of Geographers of India, Rajasthan Geographers Association, Asian Environment Council and Indian Institute of Geomorphology. He was also a Member of National Committee, Government of India for evaluating Hindi Books on environment for giving prize to the authors at National level, Ministry of Home Affairs, Government of India, New Delhi (2002). He was selected for a Research Award (Career Award) in 2002 by the University Grants Commission, New Delhi.

Prof. Kalwar is a co-ordination consultant in Geography, Vardhman Mahaveer Open University, Kota (Raj.). He has prepared 21 books for U.G., P.G. and M.Phil. students of VMOU, Kota during last two years.

Amit Kamboj did his B.Sc., M.A., Ph.D. and MCA from University of Rajasthan, Jaipur. Dr. Kamboj is a lecturer and incharge of Computer Center, Agarwal Post Graduate College, Jaipur. He has written many books of Computer Science, for the degree and diploma courses in computer for various Universities.

WATER RESOURCES AND THEIR UTILISATION

Prof. S.C. Kalwar
Dr. Amit Kamboj

Aavishkar Publishers, Distributors
Jaipur 302 003 (Raj.) India

Published by
Prem C. Bakliwal for
Aavishkar Publishers, Distributors
807, Vyas Building, Chaura Rasta
Jaipur 302 003 (Raj.) India
Phone : 0141-2578159
e-mail : aavishkarbooks@hotmail.com

2021

ISBN 978-81-7910-323-4

Price : ₹ 1800.00

Printed at
Sheetal Printers
Jaipur (Raj.) India

Preface

Water is one of the basic resources of mankind and is vital to all forms of life. Water is a biological necessity and is also of great economic significance. This being one of the basic necessities of life, Government has come out with various programmes of the supply of drinking water. The piped water supply is an economical proposition for areas with high population density. But the problem of safe drinking water assumes serious proportion when the efforts relate to the large rural populations scattered over a vast area including isolated and inaccessible habitations.

The present work is based on both primary and secondary data. To collect primary data and information from each respondent, a set of question i.e. a schedule is prepared. With the help of the schedule, relevant data and informations are collected from head of the family member through the personal interview method; Secondary data were collected from various organisations such as Public Health Engineering Department, Jaipur, Central Arid Zone Research Institute, Jodhpur, Indian Meteorology Department, Jaipur; Soil Survey Division, Jaipur, Directorate of Economics and Statistics, Jaipur; Central & State Ground Water Board, Jaipur; Irrigation Department Jaipur and Geological Survey of India, Jaipur. These data were analysed by tables and various suitable cartographic methods which are the common methods for studies of water resources.

The entire study is divided into ten chapters excluding summary and conclusion. The first two chapters are devoted to introduction and geographical profile of the study area. The remaining eight chapters deal with population, rural settlements and various aspects of water resources and management. Chapter third describes growth of population, spatio-temporal variation in population growth, distribution of rural population, impact of urbanisation on rural population, pattern of rural settlements, sex ratio, literacy, population density, occupational structure and standard of living.

Water resource potentiality and its utilization pattern i.e. surface water resources, surface water balance, hydrological cycle, subsurface geology, ground water, geomorphology, depth of water table and water potential zones have been discussed in chapter fourth while chapter fifth discusses major problem of water quality especially sources of impurities, water quality standards, water quality investigation, suitability of water for drinking purposes, causes and effects of polluted water supply, water quality control methods and major problem of fluorosis in rural Jaipur district. Next two chapters are devoted to rural water supply i.e. rural domestic water supply and water supply for agriculture. These chapters discuss rural water schemes, domestic water requirement, problems encountered in water supply in high chemical quality area, land utilization, major crops, double cropped area, irrigation water requirement and irrigated area. Chapter eight narrates impact of droughts on water supply particularly climate and related aspects, position of reservoirs in drought years, drought condition of water supply, causes of water scarcity and impact of year 1999-2002 drought. Next two chapters are very important part of the book as these chapters deal with planning and management of water resource especially current approaches, planning and programmes for future need, benefits of rural water supply, management information system in water supply, water resource management, allocation of water resources, recycling and reuse of water to meet increasing demand of water and national water supply. Last chapter is devoted to summary and conclusion.

We are grateful to Prof. H.S. Sharma, Retd. Dean, Faculty of Science, University of Rajasthan, Jaipur; Prof. R.K. Rai, Retd. Dean, Environmental Sciences, NEHU Shillong (Meghalaya) Prof. A.R. Chandrakar, Vice Chancellor, Sundar Lal Sharma Open University, Bilaspur (C.G.); Prof. Noor Mohammed, Ex-Professor and Head, Department of Geography, School of Economics, Delhi University, Delhi; Prof Santosh Shukla, Dean, Applied Sciences, H.S. Gaur

University (Central University) Sagar (M.P.), Prof. Majid Hussain, Ex-Professor and Head, Department of Geography, Jamia Millia Islamia, New Delhi for their encouragement and guidance.

We are also thankful to our colleagues Prof. M.K. Khandelwal, Prof. R.D. Gurjar, Prof. R.N. Mishra, Prof. R.D. Doi and Dr. B.L. Gupta, Department of Geography, University of Rajasthan, Jaipur for their help and co-operation.

Shri R.S. Kamboj, Mrs. Santosh Kalwar, Dr. Vidhya, Dr. Seema and Dr. Avinash Kalwar, Lecturer, Shri Bhawani Niketan Post Graduate College (Boys), Jaipur also deserve thanks for their co-operation and help.

The publisher, M/s. Aavishkar Publishers, Distributor, Jaipur has not only undertaken the task of publication but done a fairly good job in stipulated time. We are thankful to Shri Akashay ji of Aavishkar Publishers, Distributors for meticulous exceution of this publication.

Prof. S.C. Kalwar
Dr. Amit Kamboj

Contents

1

Introduction

The history of man has always been associated with water; it is well known that the earliest palaeolithic implements of human existence have been traced in the river gravels. This situation clarifies the essential need of water even for the most primitive man. While it is true that the water is the basis of all life and an absolute necessity for all, various activities like domestic, agricultural and industrial need water for which no substitute can be devised.

Water is one of the basic resources of mankind and is vital to all forms of life. Water in Indian mythology is considered as the abode of *Lord Vishnu*. One cannot imagine a man separated from water any more. Water provides a logical link between an understanding of physical and social environments. There are numerous references in the literature of people dwelling in desert to whom water is a blessing, and paradise is said to be a place where there is abundant water. Among the natural phenomena, with which man is linked geographically, almost as closely as with air, water deserves a place of first rank. It is a symbol of life. Water is a biological necessity, and is also of great economic significance. It is the supreme economic wealth, even more than the silver or gold. There is no house or human habitation in the building, of which man has not taken into consideration the proximity of water.[1]

This, being one of the basic necessities of life, Government has come out with various programmes of the supply of drinking water. The piped water supply is an economical proposition for areas with high population density. But the problem of safe drinking water assumes serious proportion when the efforts relate to the large rural population scattered over a vast area including isolated and inaccessible habitations.[2]

The issue of potable water has been attracting attention of Government and the International forum alike. The United Nations initiative in the water sector at the global level,[3] bear testimony to the interest which the Governments and agencies abroad have been taking in this regard. Back home, the *Rajiv Gandhi National Drinking Water Mission* under the auspices of the Ministry of Rural Development has been implementing the programmes of potable water supply to the population in rural areas.

In recent years increasing emphasis has been justifiably laid on the keen issues of the water supply problem. While the provision of adequate water supplies remains the key to general development in many countries, it is now recognised that the development of water resources may be planned in relation to national requirements and strategy in which the proper use of land and other important factors are fully taken into account. Furthermore, the spatial study of water in relation to man is essentially to be conceived as a geographical problem, which involves the role of such factors like climate, geology relief, drainage, rain water absorption, vegetation soils, and the man-environment relationship in general. All these aspects are more directly related to the water supply problems of rural areas, where man is primarily in contact with his rural environment than the urban areas.

As the population of the world is increasing, the demand for this vital resource is also becoming indispensable. With increasing demand of water, the scientists are paying more attention towards sources of supply, management, evaluation and utilization in general, and problems of scarcity and pollution in particular.

The pressure of demand on the limited supply has been increasing at an ominous rate as more and more people use water not only for drinking and cooking purpose but for many other purposes as well. The most acute situation is related to agriculture, which accounts for as much as 80 percent of all fresh water. Between 30 to 40 percent of the world's food production now depends on irrigation. It takes roughly 1000 tonnes of water to grow one tonne of grain and 2000 tonnes of water to grow one tonne of rice. Hence, there would be water problem whenever the demand does not adequately match with the supply. This can occur not only when there is an absolute scarcity of water as in the case of arid areas, but also when there is an excess of water as in the case of floods. The use and development of water resources may be insufficient due to inadequate domestic water supply or due to wasteful consumption, or when water quality is poor either due to natural causes or through human factors. In most of the cases several situations interact in a particular area, and jointly lead to a complex water problems. Floods can occur in very arid areas, or due to poorly drained irrigation scheme and could result in water logging of the fields and salinization of the soil and this might occur in combination with an inadequate domestic supply to the farmers. It would not be, therefore, correct to assume, that problems arise simply due to insufficiency or deterioration in the quality of water. In fact, quantity, quality and management methods are all intrinsically linked together. Furthermore, most water problems are themselves directly connected with changes and developments occurring in the human environment such as urbanization, industrialization and agricultural development. Water problems cannot, therefore, be reduced to the question of the manipulation of natural product or raw material, since water itself is a fundamental factor in shaping the environmental system in which it is used.

Usable water is limited in amount, and efforts have to be made to find out the best possible uses for the same. This will necessarily involve fixing priorities of allocation for various uses. The optimum use of water resources involves questions of conservation as well as of development and increasing attention is being given to this aspect of the problem as the demand for water grows with increasing population and greater use of land. Many surveys have accordingly been made followed by experimental and constructional works with a view to conserving a greater proportion of the rainfall by means of dams and weirs along stream courses. Moreover, absorption of rain water may be increased by appropriate methods of afforestation, and also by re-charging wells and other works, so as to effect the replenishment of the underlying ground water bodies.

1.1 SIGNIFICANCE OF THE STUDY

The natural resources provided by the earth, e.g. air, water, land, flora and fauna must be safeguarded for the benefit of present and future generations through careful technology management and planning.[4] But, as we all know, the population is increasing rapidly and becoming a burden to the limited natural resources. Due to this pressure, deterioration and depletion of cultivated land and forest, energy crisis, environmental crisis, etc. have originated. The evaluation and planning of resources is the only solution to multipronged human problems.[5]

This is specially true for Rajasthan where the nature has been very unkind in matters of availability of water. The State has only 1 percent of the water resource of the country against 11 percent of the culturable land.[6] The State has to face the menace of drought and famine regularly. Therefore, it is important to manage our water resources in the most efficient manner. It is necessary for the development of the State and for raising the standard of living of the masses.[7]

Study area that has been concentrated upon is Jaipur district which has a low potentiality of water resources, craves for want of systematic planning could not have proper water distribution so far. Specific problems in Jaipur district are like excess fluoride in water in some areas, excess iron in water, salinity of the water and limited drinking water resources. These problems are difficult to tackle with conventional water treatment techniques. Therefore, there has to be elaborate scientific and technical inputs in tackling these problems particularly for operation, maintenance and technological upgradation of installations for water supply.

On the basis of these facts, the study of the water resources of Jaipur district has been taken up to prepare a detailed resources evaluation for the district, which will promote suitable and efficient use of resources for improving the living standard of the human beings. It is worthwhile to use primary and secondary data products for preparing the inventory of some of the water resources of the area for human welfare leading to distributive justice.

1.2 OBJECTIVES OF THE STUDY

The prime objective of this study is to formulate a proposal and incorporate a water supply scheme to provide safe drinking water to villages of Jaipur district. Keeping in view the objective of the study alternate sources of water are also required to be studied which may consist of supplementing the supply by utilising surface water sources wherever feasible.

- To estimate the existing water resources in respect to the water supply and water requirement and also to assess future needs for the Jaipur district.
- Human settlements in Jaipur district and their impact on water supply.
- A critical analysis of the water resources potentiality and its application (utilisation or wastage) in study areas.
- Domestic water supply and suitability of water for drinking purposes in Jaipur district.
- To find out alternative ground water sources having acceptable quality of water for drinking purposes in the vicinity of group of problem villages and to propose regional water supply schemes.
- Irrigation water requirement and their availability for agriculture.
- To estimate water-table in the region.
- To examine change in water-table in the region.
- Cartography of salinity and total dissolved solid gradients of the water sources in the study area and their classifications according to the WHO (*World Health Organisation*) standards.
- Different techniques of Defluoridation plants and Nalgonda techniques.
- Provide Rain Water Harvesting Structures wherever feasible.
- Critical analysis of the available methods of combating the issues of water pollution.
- Evaluation of the best method of water management to ensure an equitable supply and demands of water resources in all parts of the Jaipur district.

1.3 HYPOTHESIS

The following hypothesis have been tested in the present study.

- The level of water resources development of any area depends upon the livelihood pattern of the inhabitants of the area.
- Water resource development has taken place in those areas where infrastructure is already developed.
- The present planning of water resources development is

politically oriented and it is not based on resources potentialities and necessities of inhabitants.

- The level of water resources development is very low where density of population is less.

1.4 REVIEW OF LITERATURE

The main problem of the present study is to assess the condition of rural water supply. In this field very little literature is available. The first attempt to estimate the proportion of the population of developing countries adequately served by water was made by Dieterich and Handeroon[8] (1963), where they estimated that less than 10 percent of the population of 75 developing countries had piped water in their homes including both urban and rural population. This work was especially significant as it marked the beginning of the consideration of domestic water supplies on a global basis as an element of development. The World Health Organisation estimated the maximum world population living in rural area of the 90 developing countries that were surveyed. About 12 percent of this latter group were estimated to have access to safe water supplies.

The resources potentialities have been studied and recorded through conventional as well as remote sensing techniques by the specialists of various disciplines such as ecologists, geographers, hydrologists, etc. Extensive studies of the country's natural wealth and collection of enormous qualitative and quantitative information of various resources, using remote sensing technique have been done by State Remote Sensing Application Centre, Department of Science and Technology Government of Rajasthan, Jodhpur and National Remote Sensing Agency, Hyderabad (NRSA).

Water resources have been studied by hydrologists and have also attracted the attention of Indian geographers and scientists who have also studied the problems of flood and river projects. The first systematic study in our country in the field of rural water supply was done by Mazumdar[9] (1943) followed by Arora[10] (1969), Gupta and Bhave[11] (1970), Dhar[12] (1971), Kshirsagar[13] (1971), Ram Bilas[14] (1980) and Saini[15] (1996). Sharma, et al.[16] (2000) had earlier studied the rain water harvesting for ground water recharge in Maharashtra. Minn[17] (1995) has drawn attention to ground water resource and wells. Hem[18] (1970) and Joshi[19] (1992) in their study presented the chemical characteristics of natural water and analytical approach. Trivedi[20]

(1986) evaluated the impact of green revolution on Indian farmers, while Ghosh *et.al.*[21] (1995), described basic principles of water supply in rural India. There are several published articles and websites concerning the rain water harvesting on State geographic regions and district level studies.

SAC (1986) has prepared water potential map of Gujarat and Rajasthan States. Singh[22] (1976) has described the application of remote sensing for finding ground water potential zones in Rajasthan. Sahai[23] (1980) made observation of space borne satellite data for water resources management in Panchmahal district of Gujarat.

National Thematic Mapping Organisation, (NATMO) Kolkatta (1957) has prepared forest and land use maps to show types of vegetation and their distribution. NRSA (1982) has prepared multi-date forest maps of the country to find out the declining rates in all the States. "Ground Water Atlas of Rajasthan" (1999) has been compiled by the State Remote Sensing Application Centre, Department of Science and Technology Government of Rajasthan, Jodhpur in collaboration with Ground Water Department, Government of Rajasthan, Jodhpur. "Watershed Atlas of Rajasthan" (1999) has been compiled by the State Remote Sensing Application Centre, Department of Science and Technology Government of Rajasthan, Jodhpur in collaboration with Department of Watershed Development & Soil Conservation, Government of Rajasthan, Jaipur.

Several other studies have been done by different workers pertaining to quality and quantity related to rural water supply. The contribution of NEERI (National Environmental Engineering Research Institute), is most significant in this field. In order to promote and coordinate research and development policies for rural water supply, an adhoc Working Group for Rural Water Supply and Sanitation was formed in Montreal in April 1974, with the participation of the UN Secretariat, UNDP, WHO, IBRD, FAO, UNICEF, OEC and IDRC. The Group recommended a substantial reinforcement of the International Reference Centre for Community water supply in Voorburgp Netherlands, to promote and coordinate research and developmental activities throughout the world.

A perusal of literature revealed meagre information on water supply in rural areas of country, more particularly Rajasthan where availability of potable water is a major issue before the State government. Keeping this in view, I propose to study the water supply in rural areas of Jaipur district of Rajasthan. Strategy has also been formulated to regenerate and regulate the resources in such a way that the gap between consumption and supply pattern may also be reduced.

1.5 METHODOLOGY

A foresaid work plan will be based on both primary and secondary data. These will be analysed by both empirical as well as cartographic analysis which are the common methods for water quality studies.

1.5.1 Data Base

Field observations and laboratory investigations will be dealt within separate sections. In addition, some of the field observations and controlled laboratory test procedures will also be applied.

1.5.1.1 Primary Data

To collect primary data and information from each respondent, a set of question i.e. a schedule is prepared. With the help of the schedule, relevant data and information are collected from head of the family member through the personal interview method.

This schedule includes questions pertaining to demographic details, land ownership, irrigation, cropping pattern, information regarding recharging of wells, change in production and activity, change in income. This way, we have collected necessary data and have analysed it in a proper way.

1.5.1.2 Secondary Data

Besides the landsat data, secondary data which can't be obtained through "the personal interview method" were also collected from various organisations such as Public Heath Engineering Department (PHED), Jaipur; Central Arid Zone Research Institute (CAZRI), Jodhpur; Department of Meteorology, Jaipur; Soil Survey Division, Jaipur; Directorate of Economic & Statistics, Jaipur; Central & State Ground Water Board, Jaipur; Irrigation Department, Jaipur and Geological Survey of India, Jaipur.

1.5.1.3 Method of Data Analysis

Collected data and information are arranged into a proper and appropriate tabulation format according to objectives and hypothesis of the study, so that proper conclusions can be drawn. Tables have been prepared for total value, value per household and per-capita value. It is decided to make an inter-district comparison of different variables, percentage are also calculated.

1.6 STUDY AREA

Jaipur district is one of the 32 districts of Rajasthan. It is situated between 26° 25′ to 27° 51′ north latitude and 74° 55′ to 76 10′ east longitudes. The "Jaipur district" forms the boundary with Alwar, Dausa, Tonk, Ajmer, Nagaur, Sikar district of Rajasthan and also touches the state boundary of Haryana. Jaipur district is comprised of 4 sub-divisions, 13 tehsils, 13 development blocks, 16 towns and 2187 villages. The geological formation of the region exhibits mostly of quartzite's and dolomitic marble of the Precambrian era. The region is broadly a part of Aravalli origianic system which has existed as a stable mass in the history of the earth.

1.7 STRATEGY FOR STUDY

Jaipur district is frequented by droughts which further deteriorates the availability of the ground water. Long term perspective has to be kept in view as well as the total problem of drinking water in the Jaipur district has to be viewed in its totality. In comparison to the limited availability of ground water the additional needs of the district are quite large and, therefore, priorities have to be allocated to various competitive needs.

Jaipur district is identified as having one of the worst affected district in respect of endemicity of fluorosis disease in Rajasthan. The major source of water in the rural villages of Rajasthan is the ground water. It contains a high concentration of fluorides. Though persistent efforts for improving the quality of ground water by way of resorting to defluoridation techniques have been made for various water supply schemes but the problem is still persisting, due to a rise in the use of ground water, which can be attributed to the population explosion. In addition, recharge of ground water vis-a-vis the total usage is minimal, owing to the scanty rainfall. In order to meet the requirement of water for various competitive usages the withdrawal from the ground water source is on an increase year by year whereby the availability of water is restricted to the rainfall.

The other requirements of water viz for irrigation, mining, power generation become secondary. Highest priority of water, therefore, has to be allocated for drinking purpose. For any economic activity the input of human resources is primary and, therefore, for the healthy growth of people of the district, safe and wholesome drinking water is required to be supplied.

In order to develop the water resource of the district in the scientific manner, it is necessary to carry out water balance studies for the district. At the same time it is also necessary to determine problem areas with regard to the availability of safe and potable drinking water. The needs of drinking water for these problem areas will have to be met with, by either creating additional sources of water, using the limited resources by adopting efficient scientific principles or by recycling water again after it gets polluted. After its first use, the existing available resources of water have to be reused and the possibility of diverting water from surplus areas to deficient areas have to be worked out. So far there has been too much dependence on ground water for meeting the drinking water requirements of the district.

In the rural areas the supply of water is almost entirely from ground water sources. Since the draw of water is quite excessive in this region it is suggested that for meeting the future need of drinking water source for the district, new surfaces source of water will have to be found out and then a strategy of making conjunctive use of surface and sub surface water sources has to be worked out. In order to determine the location of the possible sites for construction of dams and storing surface water it is necessary to carry out water balance studies for the region and also to impose restrictions on further exploitation of the State's water resource in a regulated manner.

In areas of scanty rainfall or where the quality of ground water is unacceptable for drinking purposes methods should be evolved to efficiently tap the micro catchments by carrying out statistical analysis on rainfall and run off as well as data to prepare suitable mathematical models for determining correct locations of water harvesting structures. The conservation of water in such locations assumes a great importance by evolving methods efficiently showing the water and if necessary to transfer it in the underground structures from where it can be easily recovered. In this context treatment of water before it can be transferred to underground structures becomes unavoidable. For this purpose it is necessary to introduce necessary legislation duly supported by an effective institutional arrangement. A well thought out policy for the management of water resource has to be spelled out under which the future needs of the drinking water should be pre-empted by making suitable reservation for this purpose.

Water has always been used and reused by man. In nature also if we look on the hydrological cycle, it consists of evaporation and precipitation signifying that it works on the principle of water reuse. It is imperative for this reason to reuse its water resource. Water supply use is not a consumptive use of water but 80 percent of the water supplied for the drinking purpose returns back as waste water and is available for reuse. It is on account of this fact that it becomes imperative for water users to accept that need of drinking water should be pre-empted. In harnessing the natural resources of water the attempt should be to adopt purely a professional approach and to induct technologies and expertise suitable for that purpose.

1.8 FRAMEWORK

The present work has been divided into ten chapters. In the first chapter introduction to water supply in rural areas and increasing emphasis has been justifiably laid on the keen issues of the water supply problem. In second chapter (Geographical Setting) presents the geographical setting of the district and examines the various variables helpful in the assessment of water resources in general.

Third chapter discusses population, occupational structure, development and pattern of rural settlement of the district. The average population growth rate in 2001 was 35.10 per cent in the district.

Water resource potentiality and its utilisation pattern has been discussed in the chapter fourth. This includes evaluation of surface and ground water resources, and its potentiality and utilization by different means.

Problem of water quality in study area has been discussed in chapter five. Major problem of fluoride, water-borne diseases and defluoride techniques in study area have been duly considered in this chapter.

Chapter sixth (Rural Domestic Water Supply) presents the utilization of water for domestic purposes which includes the detailed survey of the sources of domestic water supply for human population as well as for livestock.

Chapter seventh (Water Supply For Agriculture) includes land use, cropping pattern, cropping intensity, position of irrigation tanks and irrigation water requirement of the study area.

The impact of droughts on water supply is presented in chapter eighth. The effects of drought in sample villages have been surveyed, Reports by Government agencies were also assessed, and possible remedial measures have been suggested.

Chapter ninth deals with planning the rural water supply in the district. Different types of rural water supply programmes have been discussed including pipe water supply which has its own benefits.

Chapter tenth—Planning and Management of Water Resource comprises various issues of the planning and management of the water resource in Jaipur district. It also attempts to project the water needs for various uses in 2011 and suggests ways and means to meet these requirements. The study is concluded by a summary of the observations and findings.

1.9 NEEDS OF WATER FOR USE ON PRIORITY

Alongwith the requirement of water for domestic use and for the use of the livestock, the other priority needs of water are for irrigation, industries and thermal power. For the economic development of a region, the irrigation and the industrialisation is almost inevitable. Alongwith the needs of drinking water, the needs for water for irrigation, industry and for generation of thermal power become essential. For the future needs of water in Jaipur district more reliance will have to be given on diverting water from certain surplus areas. This will require transmission of water for sufficiently long distances and, therefore, would entail loss of water in transit and would prove to be an extra bout of expense on the State government. The total priority requirement of water will, therefore, include all the above components.

REFERENCES

1. Bhati, V.M. (1996) : *Water Resources and Their Utilisation,* Vigyan Prakashan, Jodhpur, p. 1.
2. Ghosh, G. *et.al.* (1995) : *Water Supply In Rural India,* Ashish Publishing House, New Delhi, p.ix.
3. United Nations (1977) : *Vancouver Habitat 1977 Conference, International Drinking Water and Sanitation decade programme,* UN Resolution regarding Safe Water by 2000 AD etc.

4. Stringer, E. et.al (1966) : *Geography of Resources,* Canvassel & Co., London, p. 25.

5. Kalwar, S.C. *et.al.* (1991) : *Remote Sensing in Resource Geography,* Pointer Publisher, Jaipur, p. 7.

6. Report (1988) : *Drinking Water Supply in Rajasthan,* PHED, Govt. of Rajasthan, Jaipur. p. 2.

7. Report (1997) : *Overview - Drinking Water,* PHED, Govt. of Rajasthan, Jaipur. p.8.

8. Dieterich, B.H. *et.al.* (1963) : *Urban Water Supply Condition and Needs in Seventy Five Developing Countries,* WHO, Public Health p. 23.

9. Majumdar, S.C. (1943) : *Rivers of Bengal Delta,* Associate Book Agency, Patna, p. 112.

10. Arora, S.P. (1969) : "Planning a Rural Water Supply", *The Journal of Inst. of Engg. (India),* PH2, Vol.49, No.6.

11. Gupta, V.P. *et.al.* (1970) : "A Study of Domestic Water Consumption in Nagpur", *The Journal of Inst. of Engg. (India),* Vol.51, PH1, p. 5-7.

12. Dhar, G.M. *et.al.* (1971) : "A Study of Drinking Water in Village Rahimabad of Lucknow District", *Indian Journal of Med. Res.,* Vol.59, pp. 1922-1931.

13. Kshirsagar, S.R. (1971) : "Rural Water Supply in India", *The Journal of Inst. of Engg. (India),* Vol.53, PH1, pp. 92-94.

14. Bilas, R. (1980) : "Ground Water Resources of Varanasi District, India: An Assessment of Conditions, Use and Quality", *National Geog. Journal India,* Vol.26, p.81-93.

15. Saini, S.K. (1996) : "Urban - Rural Water Supply in India", *Water Resources and Their Utilisation,* Vigyan Prakashan, Jodhpur, p. 147-170.

16. Sharma, S.K. *et.al.* (2000) : "Rain Water Harvesting For Ground Water Recharge In Maharashtra, *Ground Water Management Strategies In Arid and Semi Arid Regions,* National Seminar, Ground Water Dept. Govt. of Rajasthan, pp. 21-34.

17. Minn (1995) : *Ground Water and Wells,* Johnson Division, UOP. Inc., Saint Paul, USA.

18. Hem, J.D. (1970) : "Study and Interpretation of the Chemical Characteristics of Natural Water", *U.S. Geological Survey Water Supply,* p.1473.

19. Joshi, B.H. (1992) : *An Analytical Approach to problems of Indian Agriculture : A Theoretical and System Approach,* B.R. Publishing Corporation, Delhi.

20. Trvedi, H.K. *et.al.* (1986) : *Impact of Green Revolution on Indian Farmers,* Deep Publications, New Delhi.

21. Ghosh, G. *et.al.* (1995) : *Water Supply In Rural India,* Ashish Publishing House, New Delhi.

22. Singh, S. (I976) : "A Note on the Photographic Technique for Location Ground Water Resources in Rajasthan Desert, India," *Geographical Journal of India,* Vol. XVII, p. 107.

23. Sahai, B. *et.al.* (1980) : "Remote Sensing of Water Resources in Panchmahal district," in *Contribution of Space Observation to Water Resources Management,* Pergamon Press, Oxford, p. 49.

❑❑❑

2

Geographical Setting

2.1 INTRODUCTION

Royal land, par excellence, cradle of chivalry, *Rajasthan*, erst known as *Rajwara, Rajthana* or *Rajputana*, wears its name well as the "Land of Rajas", the "Land of the Sons of King"[1]. It is indeed no easy task to describe Rajasthan in all its vigour and liveliness with the prestige of its people whose history has so often developed at the pace of the Rajput cavalry at full gallop ! Rajasthan is, at one and the same time a great verse-chronicle of heroic exploits, an epic of immense proportions, a love story, an unceasing spiritual quest[2]. It is the adventure of a race of superb conquerors, extraordinary builders and sublime mystics whose past vanishes into the constellations of legend, often told to lure the inquisitive.

To approach Rajasthan is to embrace dozens of centuries, and a large part of the history of India itself, including some of its most glorious and sometimes tragic pages. Indeed, Rajasthan is one of the permanent bastions of Indian culture and spirituality, an essential part of India's wealth. It is a land whose people have expanded exemplary energy to safeguard its identity, its traditions, its civilisation and its very soul, in the face of the great challenge of history a magnificent example of strength and character, of bravery and sense of duty.

After independence, the State of Rajasthan was formed in 1956 with Jaipur as its capital. The history of Jaipur goes back to 1150 A.D.

when Amber was wrested from Mina chief of Susawat clan by one of the successors of Dhula Rai of Dausa. He made it the capital of the region popularly known as "Duhundhar" which remained as such for nearly six centuries. Later on, as history tells it Dhula Rai married the sister of Prithviraj Chauhan, the last chivalrous hindu king of Delhi. Later in the fourteenth century the region presently called "Shekhawati" also came into the possession of the kachhwahas. In 1660 A.D. finding the citadel of Amber Fort in which Maharaja Sawai Jai Singh-II lived, too small for his ambitious plan to expand it, he asked his Brahman Architect Vidyadhar Bahttacharya to draw up a plan and select the spot where the city was to be build. The city was subsequently founded on 18th November, 1727 and was to bear the name of Jaipur after its founder[3].

As per the signed accords and covenants the Britishers appointed political Agents and "Residents" in big princely States and established unity, order and peace in the region of Rajputana. They tried to reform and modernise the administration on the lines found useful in the adjoining British provinces. English education and social reforms were started. At this juncture attempt was made for independence by launching the first war of independence in 1857 A.D. which however, was still born, which led to the take-over of the governance of India by the British Parliament and the Queen's proclamation of the assurances of a new era of rapid progress was ushered in. University came into being, a wide net work of railways and telegraph lines were laid. A group of social reformers came forth who brought about a political awakening and the birth of Indian National Congress in 1885 A.D.

When the former Jaipur State was merged into the State of Rajasthan its was split up for administrative purposes into four districts i.e. Jaipur, Sawai Madhopur, Sikar, Jhunjhunu. But subsequently after the reorganisation of States it went on to form a part of Ajmer district.

2.2 POSITION AND EXTENT

Rajasthan is situated in the north-western part of the Indian Union (23° 30' and 30° 11' North latitude and 69° 29' and 78° 17' East longitude). The total area of the state is 3,42,239 sq. km. and ranks first in terms of area in the country[4]. It occupies 10.41 percent of the country's total area yet its population is only 5 percent of the total population of the country. The shape of the State is like a irregular rhomboid, covering a distance of 869 km. from west to east and 826 km. from north to south.

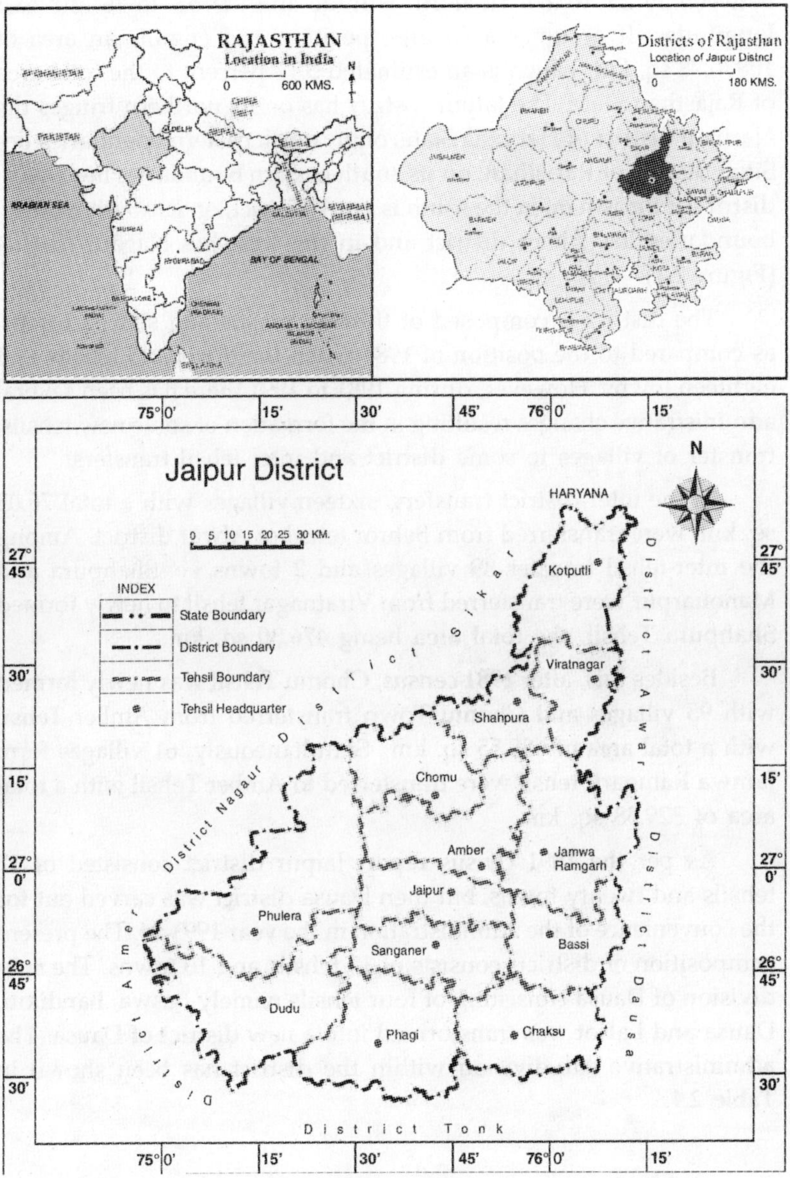

Figure : 2.1

Jaipur district the largest of its kind in Rajasthan State is one of the oldest existing districts and was carved out by portion of the erstwhile Jaipur and Kishangarh States. Jaipur district is situated between 26 25′ to 27° 51′ north latitude and 74° 55′ to 76° 10′ east longitudes. It occupies a coveted position and engulfs an area of 10876.78 sq. km. which is an estimated 3.17 percent to the total area of Rajasthan State. The Jaipur district has on its northern fringes the Narnaul district of Haryana State, on its north eastern boundaries lies Sikar district of Rajasthan, on its south eastern boundaries lies Dausa district of Rajasthan, in the south is Tonk district, on its south western boundaries lies Ajmer district and in the west lies Nagaur district (Figure 2.1).

The district is composed of thirteen tehsils and sixteen towns, as compared to the position of 1981 when there were 15 tehsils and eighteen towns. However, during 1981 to 1991 there has been several administrative changes resulting in the formation of some new tehsils, transfer of villages to some district and inter-tehsil transfers.

In the inter-district transfers, sixteen villages with a total 76.05 sq. km. were transferred from Behror tehsil of Alwar district. Among the inter-tehsil transfer 89 villages and 2 towns i.e. Shahpura and Manoharpur were transferred from Viratnagar tehsil to newly formed Shahpura Tehsil, the total area being 476.30 sq. km..

Besides this, after 1981 census, Chomu Tehsil was newly formed with 95 villages and Chomu Town transferred from Amber Tehsil with a total area of 688.55 sq. km.. Simultaneously, 61 villages form Jamwa Ramgarh tehsil were transferred to Amber Tehsil with a total area of 329.38 sq. km..

As per the 1991 census report Jaipur district consisted of 17 tehsils and twenty towns, but then Dausa district was carved out for the convenience of the administration in the year 1993-94. The present composition of districts consists of 13 tehsils and 16 towns. The sub-division of Dausa consisting of four tehsils namely Baswa, Bandikui, Dausa and Lalsot was transformed into a new district of Dausa. The administrative sub-division within the district has been shown in Table 2.1.

Table 2.1 : The Administrative Sub-division of Jaipur District (2001)

Sub Division	Tehsils	Towns	Villages	Population (2001)
1. Amber	1. Amber	1	196	261906
	2. Chomu	1	95	326304
	3. J. Ramgarh	-	241	250206
2. Sambhar	1. Phulera	4	215	389304
	2. Dudu	1	147	174683
	3. Phagi	1	170	161548
3. Jaipur	1. Jaipur	1	102	2413279
	2. Sanganer	2	209	153368
	3. Bassi	1	214	229969
	4. Chaksu	1	279	190225
4. Kotputli	1. Kotputli	1	137	339791
	2. Viratnagar	1	93	135661
	3. Shahpura	1	89	226144

Source : Directorate of Economics & Statistics, Rajasthan, Jaipur.

As per different government departments and Panchayat Samities, the basic administrative units of development, administration and planning, were listed out separately for the first time in 1981 census in relation to those districts wherein they were not coterminous with their tehsils.

The Panchayat Samities and administrative block within the district have been shown in Table 2.2.

2.3 RELIEF

The Jaipur district consists of fairly open undulating plain with hillocks in the north and north eastern part. The main peaks in the district are *Jaigarh* (648 meters), *Nahargarh* (559 meters), *Manoharpura* (747 meters), and *Oichun* (656 meters).

Table 2.2 : Different Tehsils and Blocks of Jaipur District (2001)

No.	Panchayat Samiti	Name of Tehsil it covers
1.	Bassi	Complete Bassi tehsil
2.	Sanganer	Complete Sanganer tehsil
3.	Jhotwara	Complete Jaipur tehsil
4.	Phagi	Complete Phagi tehsil
5.	Jobner	Part of Phulera tehsil
6.	Dudu	Complete Dudu tehsil and part of Phulera tehsil
7.	Govindgarh	Part of Amber tehsil
8.	Amber	Part of Amber tehsil and Jamwa Ramgarh tehsils
9.	Jamwa Ramgarh	Part of Jamwa Ramgarh tehsil
10.	Shahpura	Part of Viratnagar tehsil
11.	Viratnagar	Part of Viratnagar tehsil and Kotputli tehsil
12.	Kotputli	Part of Kotputli tehsil
13.	Chaksu	Complete Chaksu tehsil

Source : Panchayati Raj Ke Pramukh Tathya, Samudayik Vikash and Panchayat Vibhag, Rajasthan, Jaipur.

The area to the east and north of Jaipur is occupied by hill ranges and numerous isolated peaks rising up to 445 meters above mean sea level. These hills and peaks belong to the Aravalli hill range and are known by different names at different places. The longest range starting from Sambhar lake in this district crosses over up to Singhana in district Jhunjhunu. In Jaipur tehsil the range is called Puranghat and Nahargarh; in Kotputli, Viratnagar and Jamwa Ramgarh tehsils the names attributed are Ada Doongar, Chapa, Khan Rahori, Khan Dagota, Jaroonda, Khan Raipur and Khan Badri. Torawati hills stand west of rivers Sabi and Banganga. The district is drained by a number of largely non-perennial rivers of which Banganga and Sabi are the important ones. The Banganga has been impounded near Jamwa Ramgarh which provides a major share of the drinking water supplied to Jaipur city.

A contour of 381 metres (1250 feet) divides the district into two parts. The southern part is almost peneplain and the north eastern and south eastern parts are below 305 meters (1000 feet) from mean sea level. The western part lies between 381 to 533 metres (1250 feet to 1750 feet) up to a thickness of 3 to 6 metres. The remaining northern part has numerous hill ranges up to the height of 450 meters. The hill ranges run from south to north-eastern direction.

On the basis of landsite mosaic, the following hill systems are found in the Jaipur district:

1. ***Bishangarh - Baneti Range*** is from Manoharpura in Viratnagar tehsil to Baneti in Kotputli tehsil consisting Kairli, Antela, Baneti, Devipura, Balupura and Chatarpura hills.
2. ***Puranghat - Nahargarh Range*** is from Jhalana to Amarsar in Viratnagar tehsil, consisting of Jhalana, Ambagarh, Samod, Nahar and Bandol hills.
3. ***Banera - Talu Range*** is from Banera in Bassi tehsil to Talu in Viratnagar tehsil consisting Pachhodala, Khan Raipura, Khan Bhankhri, Khan Rahori, Dogota, Bamanwati and Mahnot hills.

The altitude of relief ranges from 381 to 457 meters around Jaipur, from 305 to 381 meters in the south western parts of Jaipur district, from 381 to 457 meter in the north western parts of district, from 305 to 457 meters towards north of district, from 305 to 381 meter in the eastern part of Jaipur district and from 183 to 305 meters in the south-eastern parts of the Jaipur district (Figure 2.2). In the north, north-eastern and south-eastern parts of the district, the terrain is characterised by structurally controlled NNE-SSW trending ridges comprising of Delhi quartzite. The physiography is mainly controlled by lithology and structure of the area. The common landform are escarpments as well as hogback and flat hill tops. The high dips of rocks are responsible for the formulation of escarpments. The Nidhar-Hermada hill range from the south-western extension of the main Jaipur-Shahpura hill range, running in NNW-SSE direction from water divide between Bundi and Dular, Sabi rivers. Hill features are also seen in the north-western parts of the district near Kohasina and Sankhun[5].

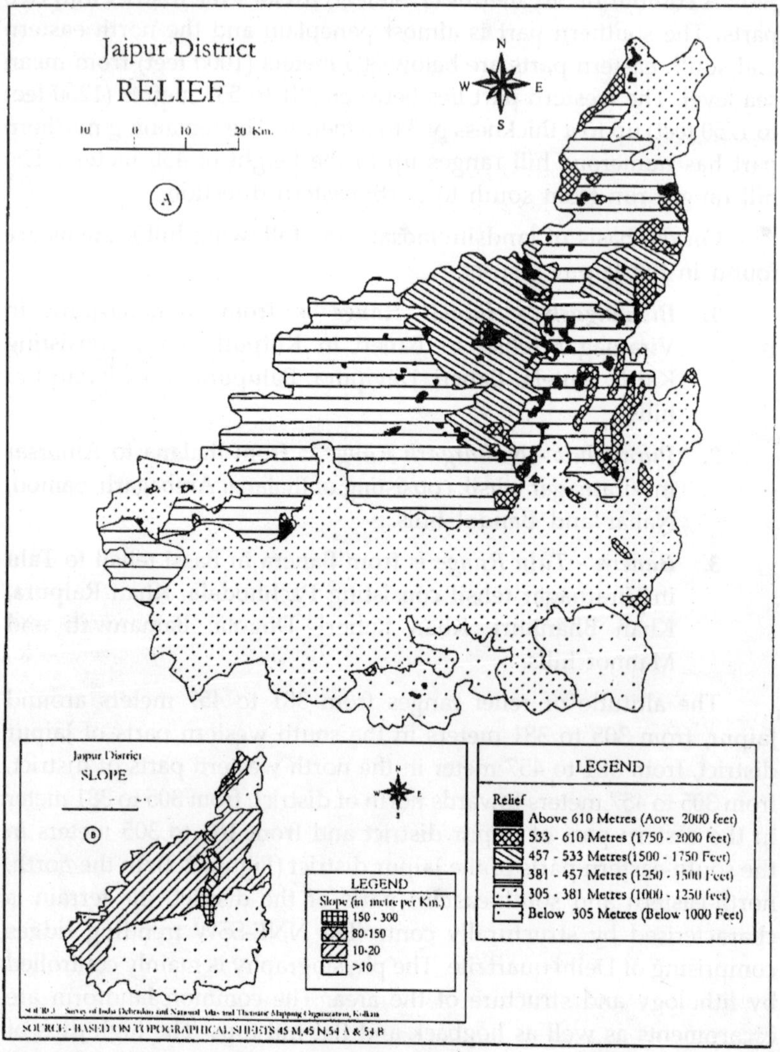

Figure : 2.2

2.4 SLOPES

The district in general is characterised by lowland topography with 0.1 percent slope. The hills and uplands have slope of 15 to 20 percent present. The area is characterised by sheet and gully erosion of moderate to severe order and the north-north western part of the district is characterised by wind erosion of moderate order.

The slope analysis plays a significant role in the planning of hill area development. The Government of India initiated the Hill Area Development Programme (HADP) in 1978. In the Rajasthan State the Remote Sensing Department, Jodhpur and Birla Science and Technology Centre, Jaipur has delineated the slopes in the Aravalli hills and developed block maps to show the hill area of various types in Jaipur district :

2.4.1 Mini Hill Tract

Mini hill tracts are scattered over Puranghat-Nahargarh hill ranges in the centre of the district, Bishangarh-Baneti hill range in the northern part of the district and Banera-Talu hill range in the central part of the Jaipur district. Table 2.3 gives an idea of the block wise mini hill tract area in the district.

Table 2.3 : Mini Hill Tract in Jaipur District

S.No.	Bolck	Area of Block in sq. km.	Hill Area in sq. km. having 30 percent or more slope	Percentage to Total Area of Block
1	Jhotwara	387.56	7.60	1.96
2	Kotputli	589.70	58.97	9.77
3	Viratnagar	674.10	80.83	11.9
4	Amber	882.64	83.17	9.42
5	Jamwa Ramgarh	1029.79	92.21	8.97
	Total	3563.79	322.78	9.05

Source : Birla Science and Technology Centre, Jaipur.

This geographical compact area of 100 sq. km. Or more with an average slope of 30 percent or more may be designated as mini hill tract for the purpose of hill development programme.

2.4.2 Small Hill Enclave

Small hill enclave is found mainly in Viratnagar and Shahpura blocks. This category covers 13.18 sq. km area in the Jaipur district (Table 2.4). This compact hill area, having 30 percent or more slopes with less than 100 sq. km. of area should be designated as small hill enclave.

Table 2.4 : Small Hill Tract in Jaipur District

S.No.	Bolck	Area of Block in sq. km.	Hill Area in sq. km. having 30 percent slopes with less than 100 sq. km area	Percentage to Total Area of Block
1	Shahpura	406.69	11.60	2.85
2	Viratnagar	674.10	1.58	0.25
	Total	**1080.79**	**13.18**	

Source : Birla, Science and Technology Centre, Jaipur.

2.4.3 Other Hill Area

Other Hill areas have less than 30 percent slope. This type of area requires steps for raising resource base in the form of fuel wood plantation, pasture development, silvi pastoral plantation, sand dune stabilisation etc. This type of slope area is extended at all over the foot hills of the three hill ranges in the district. Total 1001.64 sq. km area is under this category (Table 2.5).

Table 2.5 : Other Hill Areas in Jaipur District

S.No.	Bolck	Area of Block in sq. km.	Hill Area in sq. km. having less than 30 percent slope	Percentage to Total Area of Block
1	Jhotwara	387.56	32.52	8.41
2	Shahpura	406.69	78.70	19.35
3	Kotputli	589.71	63.61	10.79
4	Bassi	630.21	50.50	9.23
5	Viratnagar	674.14	230.75	34.23
6	Amber	882.64	224.54	25.43
7	Jamwa Ramgarh	1027.70	297.45	28.94
8	Dudu	1857.34	23.57	1.26
	Total	6455.99	1001.64	15.51

Source : Birla, Science and Technology Centre, Jaipur.

2.5 GEOLOGICAL RESOURCES

The hydrological frame work of the district is essentially controlled by geological setting as circulation and movement of ground water is controlled by the interconnected primary and secondary porosities of the geological formation constituting the aquifers.

A large part of the district is covered by a thick mantle of soil, blown sand and alluvium. The oldest rocks in Jaipur district are the schists gneisses, migmatites and quartzite's of pre Aravalli region which are considered to be nearly 2,500 million years old. These rocks are mostly covered under a mantle of sand and alluvium of Recent to sub-Recent age. Overlying these rocks, with a major unconformity, are the rocks of Delhi super group, which is made up of Raialo, Alwar and Ajabgarh group.

The general stratigraphical succession in the district after Geological Survey of India is as follows and has been depicted in Table 2.6.

Table 2.6 : Stratigraphical Succession in Jaipur District

Era	Period	Rank	Lithology
Quaternary	Recent to sub-Recent	Alluvium	Alluvium, soil, sand, dune sand, river gravels and terraces etc.
	Post-Delhi		Granite, quartz veins, pegmatites, amphibolite (Instrusives).
Proterozoic	Delhi Super Group	Ajabgarh Group	Schists, phyllites, marble and quartzites.
		Alwar Group	Quartzites, conglomerate and schists.
		Raialo Group	Dolomitic marble and quartizte.
Arehaean	Bhilwara Super Group		Genesisses, schists and migmatites.

Source : Geological Survey of India, Jaipur.

2.5.1 Delhi Super Group

The rocks of Delhi super group form high relief in the north and northeast part of district. The rocks of Delhi super group, consist of Ajabgarh, Alwar and Raialo group (Figure 2.3).

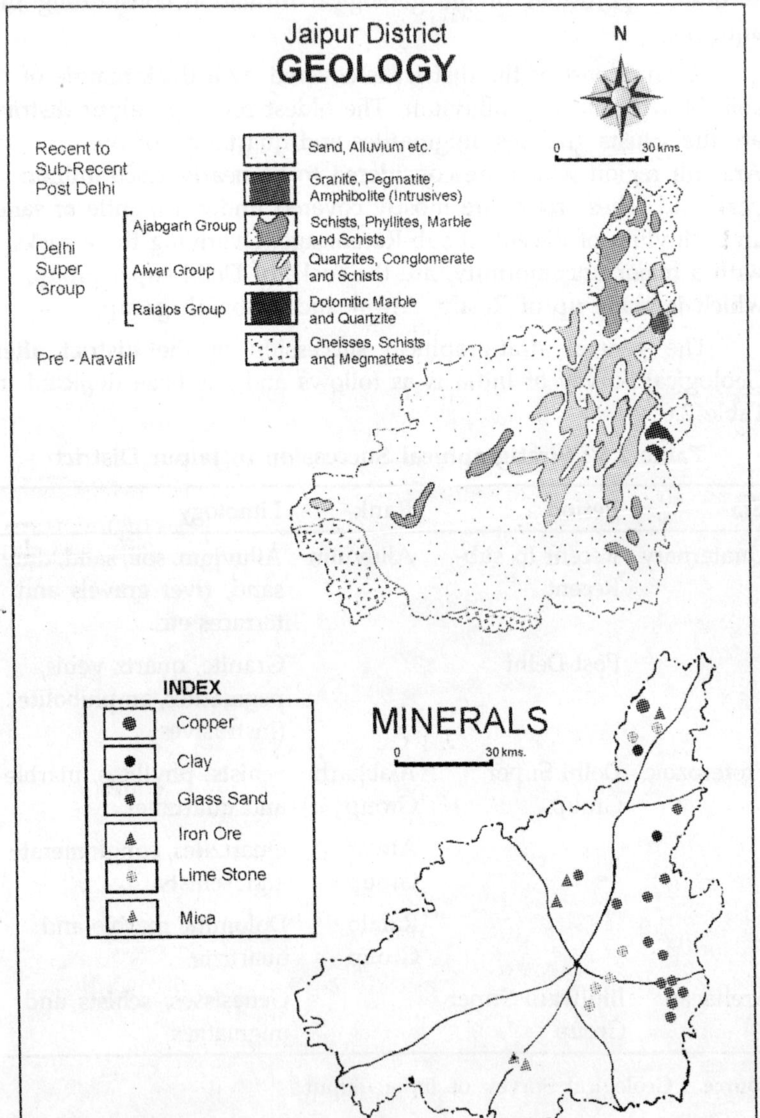

Figure : 2.3

2.5.1.1 Ajabgarh Group

The Ajabgarh formation is predominantly, argillaceous in composition, comprising slates, phyllites, garnetiferous biotitie, schist, with bands of quartzite and marble. Outcrops of Ajabgarh are seen towards north of Shahpura and west of Kotputli, near Antela, Begrana, Dantil, Kilanpur, Mehranpur, Pawana and Sarand. The major folds of the area plunge 40° - 50° towards northeast. The regional strike in the area is NNE -SSW (north to north east and south to south west) with dips varying from 45° - 70° towards WNW and ESE.

2.5.1.2 Alwar Group

Rocks of Alwar group are predominantly arenaceous in composition and comprise of orthoquartzite, micaceous quartzite and ferruginous quartzite with subordinate bands of quartz schist and quartz biotite schist. These rocks are exposed as high ridges near Hermada, Samod, Amber, Jobner, Bachun, Chomu, around Jamwa Ramgarh and near Bairath. From north Jaipur city a long but discontinuous ridge of quartzite extends right up to Shahpura and beyond in a northeast to southwest direction. Quartzite's are commonly grey but are also found in different colours such as fawn, buff and white. These are medium grained and ranges between felspathic grit to sericitic quartzite. The gritty quartzite are pink and coarse grained and sericitic quartzite are medium grained and greyish to white in colour.

2.5.1.3 Raialo Group

The rocks of Raialo group comprise of mainly dolomite marble and minor quartzite's. The Raialos are exposed in a small patch towards northeast of Andhi, in north east part of the district.

2.5.2 Post Delhi Intrusive

The post Delhi intrusive of the area are amphibolites, granite, pegmatite and vein quartz. Amphibolites is found as intrusive in quartzose and schistose rocks in Viratnagar valley around Rampur and is found in few well section close to Kalwar and Ballukhan-ki-Dhani. The granite occurs as bosses within Delhi in the Viratnagar Valley near Bawli-Ka-Bas and Pahari. Granites of post-Delhi are exposed around Sankkur and Rahlana in south western part of the district. Veins of pegmatite and quartz traverse the Pre-Aravalli and Delhi rocks in the whole of the area.

2.5.3 Bhilwara Super Group

The Bhilwara super group (more than 2,500 million years old) of rocks comprising mica-schist's and gneisses occupies major part of the district, i.e. Dudu, Sambhar[6]. The area immediately to the south of Sambhar lake is schistose, while the area to the south of Dudu is gneissic. The foliation trend is along NNE-SSW (north to north east and south to south west) has a steep dip. The rocks are sparingly jointed with rectangular pattern having predominating east-west and north-south joints. In the open well sections the zone of weathering has been noticed down to 10 metres. Most of the dug wells in Dudu-Sambhar area are located in these rock units.

2.5.4 Structure

Rocks of different age groups of the district are tectonically very deformed as evidenced by numerous major and minor faults, folds, joints and fractures and major liniaments of the area have been marked in the hydrogeological map of the Jaipur district. The major structural feature in the area around Viratnagar and Kotputli is NNE - SSW trending syncline corresponding to the Sabi river valley which has been formed due to erosion of the younger Ajabgarh rocks[7].

2.6 CLIMATE

Climate, the synthesis of the weather elements, temperature, pressure, precipitation and humidity etc., are more or less remotely relevant to water supply. For example, it is clear that variations in the number of cloudless days in rainy season and in mean summer temperature affect both the demand for water deliveries, especially through increased field irrigation and the available supply in a surface storage system through surface evaporation. It is, however, the variation in precipitation that is most closely associated with drought. Thus, weather can often be a limiting factor on the specific amount of water available to a specific area.

Semiarid type of climate prevails over the district. The district has a dry climate except during the south-west monsoon season. December to February is the winter season, after which the hot season commences and continues till about the third week of June when the south west monsoon sets in. The south-west monsoon season is comparatively short in this region and lasts only till mid-

September. The period from the second half of September to the end of November is the post monsoon or retreating monsoon season. Occasional hailstorms and cyclonic rains are common during the winter months.

2.6.1 Solar Radiation

All the physical and biological processes occurring near the surface of, or in the atmosphere involve some energy transpirations. Processes such as photosynthesis on which human existence partly depends, are virtually impossible without solar radiation, which is also a significant component of the energy balance and the hydrological cycle. Most of the climatologically observation network in India does not measure the components of radiation including Jaipur observation station. At any place available radiation depends upon types of surface cover, water-bodies slope and elevation of surface. Here it is only necessary to know about the importance of solar radiation, its effect and relation with other climatic elements. Detailed study of solar radiation is beyond the scope of present study.

2.6.2 Temperature

The only meteorological observation centre in the district is located in Sanganer town. The data of this station may be known as representing the weather conditions in the district as a whole. A careful study of temperature distribution, its ranges and seasonal march provide adequate tool in assessing the climate of a region.

For the last forty years (1960 to 2000), monthly mean temperature is represented in Table 2.7. The maximum monthly temperature in the area has been recorded in May and June, which varies from 40.60°C to 39.20°C. May and the first half of June being the hottest part of the year.

The minimum monthly mean temperature is recorded in December and January, which varies from 9.10°C to 8.30°C. These months being the coldest part of the year. In the last six years the maximum temperature exceeded to 44.8°C in 1999 and 44.3°C in 2000 (Table 2.8).

During summer the plain as well as upland suffers from extreme heating and hot westerly wind known as *"Loo"*. After mid November both day and night temperatures fall rapidly till January.

Table 2.7 : Month wise Mean Rainfall, Temperatures, Potential Evaporation, Relative Humidity, Wind Speed and Dust Strom of Jaipur District (1960 to 2000)

Months	Mean Rainfall (cm)	Mean Temperature (^{0}C)	Max. Temperature (^{0}C)	Min. Temperature (^{0}C)	Potential Evaporation	Relative Humidity (%)	Wind Speed (Km/hr)	Dust Storm
January	1.12	15.15	22.40	8.30	7.56	47.50	11.00	0.00
February	0.90	18.05	25.40	10.70	9.82	38.50	11.50	0.00
March	0.59	23.20	30.90	15.50	15.27	30.00	12.00	0.30
April	0.36	28.75	36.50	21.00	19.64	24.00	13.10	0.70
May	0.99	33.20	40.60	25.80	25.74	26.00	16.50	1.70
June	5.13	33.25	39.20	27.30	24.26	40.00	17.80	3.00
July	18.27	29.85	34.10	25.60	16.17	68.50	16.20	0.10
August	18.07	28.10	31.90	24.30	12.64	76.50	12.40	0.10
September	8.52	28.10	33.20	23.80	14.48	65.50	12.30	0.00
October	0.99	25.27	33.20	18.30	13.72	41.50	11.80	0.00
November	0.19	20.50	28.80	12.80	8.47	39.00	7.60	0.10
December	0.57	6.75	24.40	9.10	6.78	45.50	8.10	0.00

Source : India Meteorological Department, Jaipur.

2.6.2.1. Range of Temperature

The diurnal range of temperature differs from month to month (Table 2.7). The study of month-wise mean maximum and mean minimum temperature represents that even the night in May and June are comparatively warmer than the days in January. The days in the month of January are somewhat pleasant. In last seven years the hottest day in the year 2001 was 47.0°C and the coldest day in the year 1998 was recorded 2.6°C. The study of last seven year wise maximum, minimum and mean temperature has been recorded in Table 2.8. In three years increase in mean daily temperature of Jaipur district during the year 1999 is 26.3°C, in year 2000 is 26.5°C and year 2001 is 26.6°C. (Table 2.8) have been recorded.

Table 2.8 : Max. Temperature, Min. Temperature, Mean Temperature, and Mean Humidity in Jaipur District (1995-2001)

(Degrees Centigrade)

Year	Maximum Temperature	Minimum Temperature	Mean Temperature	Mean Humidity
1995	46.0	4.0	25.5	58.70%
1996	43.6	4.8	25.5	51.00%
1997	43.2	3.7	24.3	55.00%
1998	46.6	2.6	25.9	52.00%
1999	44.8	4.1	26.3	47.00%
2000	44.3	2.9	26.5	43.00%
2001	47.0	3.0	26.6	41.00%

Source : India Meteorological Department, Jaipur.

2.6.3 Relative Humidity

Relative humidity over land surface in general is largely governed by the direction of wind in relation to the position of a larger water body. Relative humidity during day time varies inversely with temperature. The atmosphere is damp in the morning hours and its dryness gradually increases as the sun ascends in the sky. Again in the evening with the decreasing temperature relative humidity starts increasing.

During the brief south west monsoon the relative humidity is generally over 60 percent and in the rest of the year air is mostly dry. In the summer season which is the driest part of the year, relative humidity in the afternoon may get as low as 15 to 20 percent.

Relative humidity is quite high during the main monsoon (July to September) amounting to 68.5, 76.5 and 65.5 percent respectively (Table 2.7). August is the dampest month and relative humidity touches 80 percent in the morning hours in the month. During the remaining months, the area mostly comes under the influence of dry continental air masses and hence in these months the relative humidity is lower.

In the district, mean relative humidity is sufficiently low and it has never exceeded over 55 percent in seven years as shown in Table 2.8. Relative humidity of Jaipur district has been continually decreasing since last seven years from 1995 to 2001.

2.6.4 Winds

Wind has also a marked variation from winter to summer. The westerly winds are experienced in major part of year, whereas southerly winds have the shortest duration. Winds are generally light than moderate. Westerly to south-westerly winds prevail in the south-west monsoon season. In the post-monsoon and winter months winds are mostly from the direction between west and north. In the summer season the winds blow from south-west to north-east.

Normally the wind velocity is high in May (16.50 km/hr), June (17.80 km/hr) and July (16.20 km/hr) (Table 2.7). In the months of May and June, the wind is often very fast and sometimes records average speed of 45 km per hour as was in Sambhar and 48 km per hour as was at Jaipur. It can be concluded that the speed of storms increases from south-west to north-east.

Severe dust storms, locally known as "Andhi" and the gusty-dusty hot wind, locally known as "Loo" are among the chief weather phenomena in the area during the dry summer months. Data of Jaipur district reveals that Jaipur had 3 storms during June respectively (Table 2.7).

2.6.5 Clouds

During the south-west monsoon season the sky is moderately to heavily clouded and overcast of downpour dominates some days.

In the remaining season clear or highly clouded skies prevail occasionally, in winter season the sky becomes cloudy when the western disturbances pass over the district.

2.6.6 Rain Fall

Among all the weather elements after temperature, precipitation is the most important phenomenon because it is the ultimate source of all the surface and sub-surface water [8]. In the different forms of precipitation, it mainly occurs as rainfall in the area under study.

2.6.6.1 Intensity of precipitation

Basically, the pattern of mean rain-day follows the same pattern as for mean rainfall amount. The mean annual rain-day recorded at Jaipur district is 32 rainy days (with rain of 2.5 mm to 10 cm. or more) of which 90 percent days occur during the monsoon season, i.e., June, July, August, September. The statistics of intensity of rainfall highlight the nature of precipitation, whether it is stormy or spasmodic or evenly distributed; more over, it is also clear that monthly rainfall intensity is high in the monsoon months i.e., in June 5.13 cm., July 18.27 cm., August 18.07 cm. and September 8.52 cm. (Table 2.7) and decreases towards the both pre and post monsoon months.

2.6.6.2 Areal Distribution

The rainfall generally increases from the western part to the eastern part. But in the Amber region and Viratnagar region the rain fall is a little higher than the surrounding parts of the district due to forested hills. An Isohyetal map of the district is given in Figure 2.4.

2.6.6.3 Seasonal Distribution

The distribution of rainfall in space and time is, however, uncertain and unreliable, which are the main characteristics of the Indian monsoon. The data of commencement also varies from year to year, but it generally bursts in the last June or in early July.

Of the total annual rainfall of Jaipur district, 3.6 percent is obtained in winter season, 3.5 percent in the summer season, 89.8 percent during the summer monsoon and 3.1 percent in the post-monsoon and gradual decrease till summer season (Table 2.9).

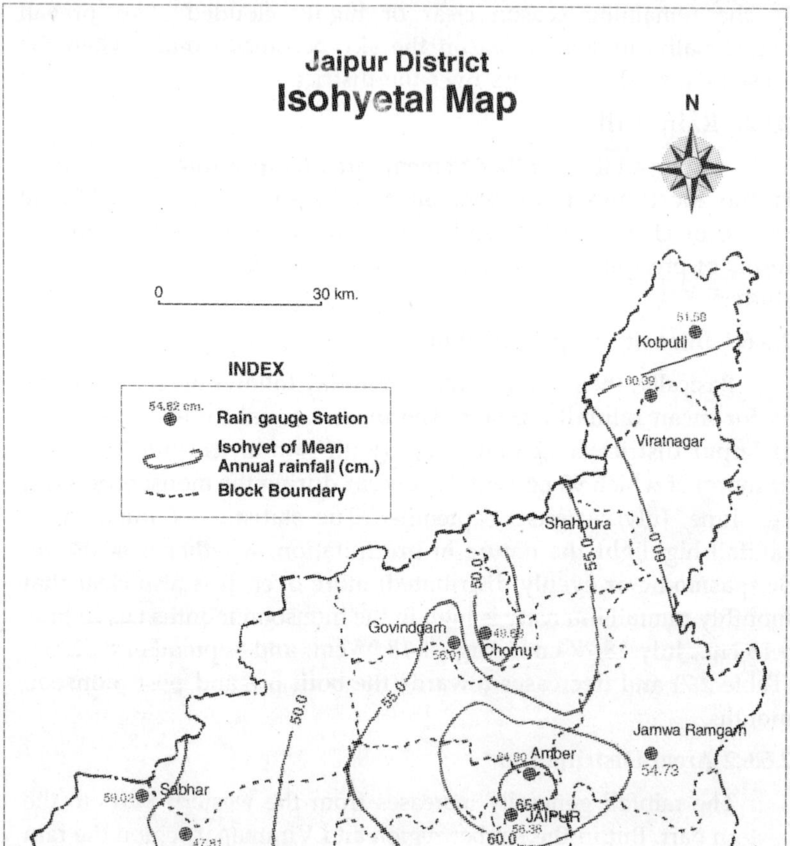

Source : Central Ground Water Board, Jaipur

Figure : 2.4

2.6.6.4 Annual variation of rainfall

The variation of the annual rainfall from year is very large. The Normal mean annual rainfall in the Jaipur district is 54.82 cm. (Table 2.10). However long term normal rain fall shows that mean annual rainfall in the district has slightly increased over the year.

Table 2.9 : Monthly Mean Rainfall and Seasonal Rainfall in percentage of Jaipur District

Season	Month	Mean Rainfall (cm)	Monthly Rainfall in percentage (%)	Seasonal Rainfall in percentage (%)
Winter	January	1.12	2.0	3.6
	February	0.90	1.6	
Summer	March	0.59	1.1	3.5
	April	0.36	0.6	
	May	0.99	1.8	
Monsoon	June	5.13	9.2	89.8
	July	18.27	32.8	
	August	18.07	32.5	
	September	8.52	15.3	
Post Monsoon	October	0.99	1.8	3.1
	November	0.19	0.3	
	December	0.57	1.0	

Source : India Meteorological Department, Jaipur.

The mean annual rainfall is more than the normal rainfall at all the stations. But last three years 1999 to 2001 (drought year) the mean actual annual rainfall is less than the normal annual rainfall (Appendix-1).

Table 2.10 : Normal Annual Rainfall, Actual Annual Rainfall and their difference for Jaipur District

Year	Normal Annual Rainfall (cm.)	Actual Annual Rainfall (cm.)	Difference with normal Increasing (+) and Decreasing (-)
1995	54.82	84.91	(+) 30.09
1996	54.82	99.53	(+) 44.71
1997	54.82	78.16	(+) 23.34
1998	54.82	63.29	(+) 8.47
1999	54.82	43.15	(-) 11.67
2000	54.82	37.59	(-) 17.23
2001	54.82	37.77	(-) 17.05

Source : India Meteorological Department, Jaipur.

2.6.6.5 Variability of Rainfall

The distribution of mean annual rainfall is uniform in different rain gauge station of Jaipur district. Among the 17 rain gauge stations (Table 2.11), mean annual rain gauge varies from 45.37 cm (Phulera) to 65.74 cm (Amber) with an average of 56.22 cm. The coefficient of variation among these rain gauge station is quite low i.e. about 10 percent of mean annual rainfall. This indicates that rainfall is quite uniform throughout the district.

Table 2.11 : Comparisons of Normal & Mean Annual Rainfall for Jaipur District

S.No.	Rangauge Station	Normal Annual Rainfall (cm)	Mean Annual Rainfall	Variation of Mean from Normal	Variation of Mean from Normal (%)
1	Jaipur	56.38	61.94	5.56	9.86
2	Bassi	49.76	57.47	7.71	15.49
3	Chaksu	58.39	61.83	3.44	5.89
4	Sanganer	57.18	59.58	2.40	4.20
5	Amber	61.80	65.74	3.94	6.38
6	J. Ramgarh	54.73	59.07	4.34	7.93
7	Chomu	52.63	56.01	3.38	6.42
8	Samod	52.27	49.43	-2.84	-5.43
9	Sambhar	47.87	50.13	2.26	4.72
10	Phulera	42.75	45.37	2.62	6.13
11	Naraina	49.37	49.69	0.32	0.65
12	Dudu	47.81	51.77	3.96	8.28
13	Phagi	48.66	53.96	5.30	10.89
14	Kotputli	51.58	56.18	4.60	8.92
15	Pawata	52.41	60.93	8.52	16.26
16	Shahpura		56.22		
17	Viratnagar	59.39	60.44	1.05	1.77

Source : India Meteorological Department, Jaipur.

2.6.6.6 Probability Analysis

Probability analysis of annual rainfall of Jaipur district has based on Chegodev's empirical relation, occurrence of amount rainfall for different probability percentages (Table 2.12). It is observed that there is 90% probability that annual rainfall will exceed 22.50 cm. at all the rain gauge stations in the district and maximum 90% probability in Viratnagar tehsil. The average probability of occurrence of mean annual rainfall in the district is around 45%.

Table 2.12 : Probability of Exceedence of Annual Rainfall for Jaipur District

No	Rain gauge	StationProbability								
		10%	20%	30%	40%	50%	60%	70%	80%	90%
1	Chomu	86.53	75.40	70.24	61.64	51.65	42.54	39.43	35.68	31.31
2	Naraina	78.58	73.86	53.67	49.50	44.46	41.29	37.64	35.49	29.69
3	Pawata	94.39	86.40	76.40	68.43	56.15	51.88	47.52	38.95	26.93
4	Viratnagar	83.39	79.54	68.13	61.30	58.12	54.31	47.79	41.12	35.40
5	Kotputli	84.78	73.69	65.89	58.63	49.78	46.27	41.55	37.19	27.59
6	Phagi	84.15	73.80	65.76	55.39	49.35	43.81	39.51	35.79	27.23
7	Dudu	82.30	66.19	59.98	58.48	50.13	42.90	38.31	32.87	30.13
8	Sambhar	78.83	70.83	57.84	51.82	44.11	41.77	38.01	26.91	22.73
9	J.Ramgarh	94.45	77.54	69.75	60.18	50.58	45.54	43.57	41.98	32.72
10	Sanganer	89.47	76.43	66.57	64.27	58.33	53.23	47.55	40.42	31.26
11	Chaksu	98.31	895.3	71.38	66.82	56.96	53.99	49.45	38.64	27.22
12	Bassi	107.4	82.13	67.63	60.95	50.31	45.28	39.34	33.65	22.50
13	Phulera	66.25	57.27	52.16	48.31	43.05	38.84	37.38	35.52	28.84
14	Amber	93.48	88.92	80.57	688.1	66.93	64.88	55.48	47.92	39.59
15	Jaipur	99.94	84.87	78.50	66.26	59.82	55.22	45.59	39.88	32.51
16	Samod	81.95	68.85	54.60	49.79	46.82	39.83	36.38	33.73	26.38

Source : India Meteorological Department, Jaipur.

2.7 SOIL

The importance of soil for an agricultural region like the study area can hardly be exaggerated. All the plant life is dependent upon the intake of water by soil. Springs, wells, groundwater supplies and the base flow of stream are dependent on it. Rajasthan, being geographically the largest state in India, has proportionally a greater soil resource[9]. The study of soil is not only important for agricultural practices but also it helps in the study of ground water condition and composition. As we know, the main source of ground recharge is the rainfall, but the soil-cover determines how much of the rain will enter the soil for replenishing the exhausted ground water reserve. The infiltration capacity of a soil depends on the texture of the soil, on its thickness and degree of compactness. Surface run-off can only occur when the intensity of the rainfall exceeds the infiltration capacity of the soil. Although different types of soil have different infiltration capacity, e.g. a clay soil composed of many fine particles and containing much interstitial water has a low infiltration capacity, while a dry sandy soil will at least for a time absorbs and pass water in greater quantity[10].

Some of the diseases causing bacteria and viruses depend on soil composition and structure for their movement. The chemical composition of water either increases or remains unaffected during movement through soil, but bacteria and viruses are mostly filtered out when passing through vertical column of soil. Its concentration depends on particular types of soil, e.g., viruses are readily absorbed on to clay under appropriate conditions, and the higher the clay content of soil, the greater is the expected removal [11]. Similarly sandy loams and soils containing organic matter are also favourable for virus removal, but soil containing sand or sand and gravel mixtures do not achieve good removal and can present serious problem of ground water pollution. So, for understanding the importance of soil in respect of water availability its condition and composition of a particular area needs detailed study.

The soils of the study area are formed due to the variation of alluvium with multiphase profiles of alternating layers of sands to clay. In appraising the productivity of the soil of an area, there is need to have a deep knowledge of the solid types and their geographical distribution. This is especially true where the soils are affected by various natural hazards like salinity and alkalinity, water logging, desertification etc. On the basis of colours and texture the soils of Jaipur district can be categorised into following seven categories (Figure 2.5).

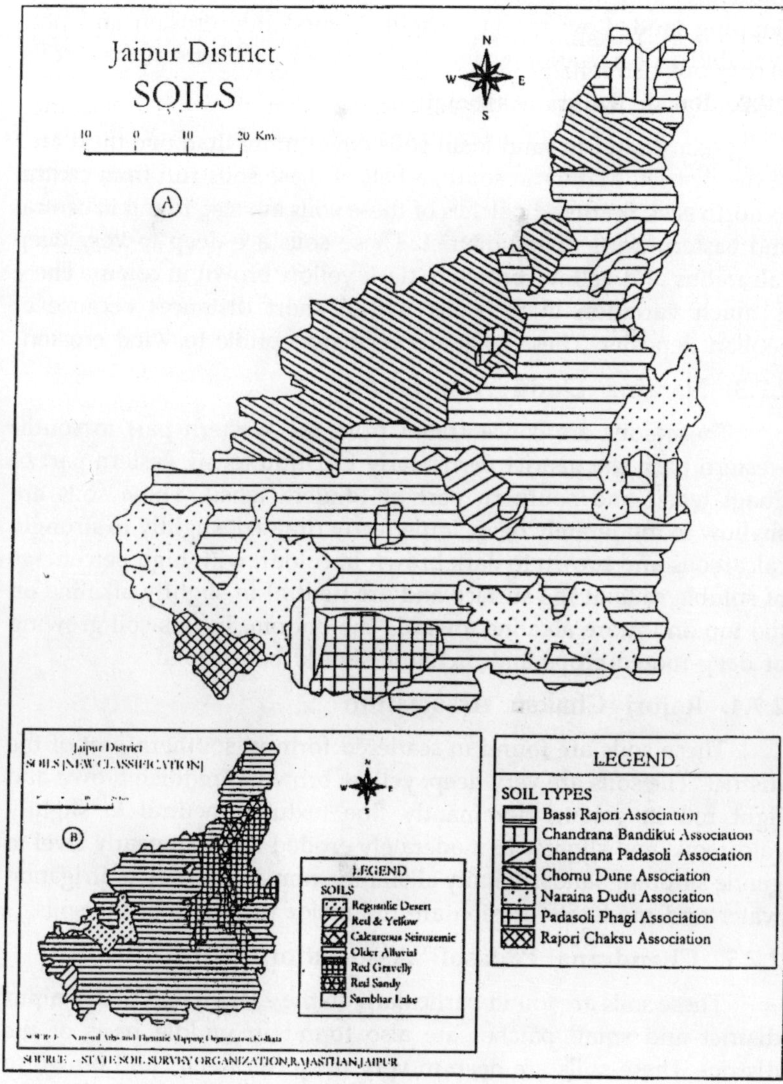

Figure : 2.5

2.7.1. Chomu Dune Association

These soils are found particularly in the north-western part of the Jaipur district and small patches are also found in western and north-eastern parts of the district. These soils are deep to very deep calcareous and moderately eroded lying on nearly level to gently slopping land. Low fertility, slight to moderate erosion and non-availability of irrigation water are the major problems of these soils.

2.7.2. Bassi Rajori Association

Loamy sand to sand loam soils cover more than one third area of the district. Mainly in south, a belt of these soils run from central to north east. Scattered patches of these soils are also found in central and eastern parts of the district. These soils are deep to very deep calcareous and yellow, brown to dark yellow brown in colour. There is much variation in the soils within short distances because of aeolian deposits. These soils are very susceptible to wind erosion.

2.7.3. Naraina Dudu Association

These soils are concentrated from the western part to south-western part, the district particularly in Dudu tehsil, eastern part of Phagi tehsil and southern part of Phulera tehsil. These soils are shallow to moderately deep, imperfectly drained, slightly to strongly calcareous and brown to dark brown in colour, with high percentage of soluble salts at the surface and are neutral to slightly alkaline on the top and more alkaline towards the bottom. In this soil growing of deep-rooted crops and orchards should be avoided.

2.7.4. Rajori Chaksu Association

These soils are found in scattered form in southern part of the district. The soils are very deep, yellow brown to reddish brown and light red in colour, dominantly fine texture neutral to slightly calcareous and slightly to moderately eroded lying on nearly level to gentle slopping lands, Salinity alkalinity, non-availability of irrigation water and moderate erosion are the major problems of the soils.

2.7.5. Chandrana Padasoli Association

These soils are found particularly in the eastern part of the Jaipur district and small patches are also found in middle parts of the district. These soils are deep to very deep calcareous and brown to dark grey in colour. the soil profile is mature with kankar interspersed between 40 to 105 cm depth. The water holding capacity is low, slightly alkaline on the top and more alkaline towards the bottom, low fertility, slight to moderate erosion and non-availability of irrigation water are the major problems of these soils.

2.7.6. Chandrana Bandikui Association

These soils are found in the eastern part of the district particularly in Bassi tehsil. The soils are very deep, yellow brown to dark grey brown in colour, dominantly fine texture neutral to slightly calcareous and slightly to moderately eroded lying on nearly level to gentle slopping lands. The soils are neutral to alkaline in the reaction, poor quality of irrigation water and low water holding capacity.

2.7.7. Padasoli Phagi Association

These soils are found in scattered form in southern part and some patches are also found in Phulera and Chomu tehsil of the district. The soils are deep, dominantly fine textured, poorly to moderately well-drained, calcareous sub solid, and slightly to moderately eroded lying on nearly level to gentle slopping lands. These group of soil is dark brown to dark grey in colour, with slight to moderate salinity, alkalinity, non-availability and poor quality of irrigation water and erosion are the principal limitation of these soils.

2.8 SOIL FERTILITY

The soils of the district are very low in available nitrogen. About 77 percent of the soils samples analysed fall into the category of very low except Shahpura block. Shahpura block has about 10% soil samples in medium category. The nutrient index value varies from 1.0 to 1.11 with an average value for the district as 1.04. The soils of the district are rich in available phosphorus as about 56% of the soil samples analysed by the Soil Survey Department are high in available phosphorus. The nutrient indices 2.41 as an average value for the district and maximum 2.74 in Viratnagar. Out of 13 blocks 10 are high in this nutrient and their ascending order is Phagi, Amber, Dudu, Shahpura, Jhotwara, Sambhar, Kotputli, Bassi, Jamwa Ramgarh and Viratnagar. Govindgarh and Chaksu fall in medium category. However, the Sanganer block has low nutrient index (1.65) as shown in Appendix-2.

The nutrient index value for the district ranges from 1.90 to 2.25, which shows that soils have medium status of potash. Bassi, Sambhar Lake and Govindgarh block have nutrient index value below 2.00, while rest of the blocks have nutrient index value above two. Maximum alkalinity problem is found in Dudu block followed by Phagi and Sanganer.

2.9 TRANSPORT AND COMMUNICATION RESOURCES

The district is well connected by rail, road and air routes. Falling on the broad & metre gauge section of the East-Western Railway Zone, Jaipur, the district headquarter, is directly linked with Delhi, Agra, Ahmedabad and with other major cities of Rajasthan. There is an international airport at Sanganer located at a distance of 11 km from the city. Air services connect Jaipur with Delhi, Agra, Mumbai, Aurangabad, Ahmedabad, Jodhpur, Udaipur and aboard. Jaipur is also connected by extensive network of roads with all the district headquarters of the State and major cities of our country.

National highway 8, 11 and 12 pass through the district. Besides Jaipur, eight other towns viz. Kotputli, Amber, Shahpura, Manoharpura, Sanganer, Dausa, Bassi and Chomu are situated on the national highways.

The road development in the district has made significant progress during the last thirty years. The district had 866 kilometres of roads in 1950-51, which increased to 4102 kilometres (Appendix-3) in 2001-02.[12] The principal categories of powered vehicles are private cars, jeeps, motor cycles and auto-rickshaw, contract taxies, public and private carriers and tractors.

REFERENCES

1. Rajvanshy, P.S. et. al. (1987) : *Study and Status of Environmental Consciousness In the Planned City Jaipur*. Rajasthan Pollution Prevention and Control Board, Jaipur. p. 1.
2. Rajvanshy, P.S. et. al. (1987) : *Ibid*, p. 2
3. Misar, B.P. et. al. (1962) : Translation of "*History of Rajasthan by Col. James Tod*", reprint in 1987, publisher unique trades, Jaipur. p.600.
4. Arya, R.P. et. al. (2000) : *Road Atlas, Rajasthan*, Indian Map Service Publication, Jodhpur. p. 3.
5. Murthy, R.S. et. al. (1979) : "Land Resources Inventory of Rajasthan by Using Colour Composites of Space Imagery", *Science Review*, CAZRI, Jodhpur.
6. Website : *http://www.rajamb.com/Geology.htm*, Rajasthan State Agriculture Marketing Board, Jaipur.
7. Siddiqui (1970) : *General Report*, Geological Survey of India. p. 76.
8. Beasdale, A. et.al. (1963) : "Study and Assessment of Water Resource", *Conservation of Water Resources*, Inst. Civil Engg. Symposium. p. 121-136.

9. Website : *http://www.rajamb.com/soils.htm,* Rajasthan State Agriculture Marketing Board, Jaipur.

10. Morgan, M.A. (1969) : "Overland Flow and Man", *Introduction to Geographical Hydrology,* Chorley, R.J. ed., Methuren & Co. Ltd, London, p. 56.

11. WHO (1979) : Human Viruses in Water, *Waste Water and Soil,* WHO Technical Report Series No. 639.

12. Abstract (2001) : *District Statistics Abstract,* Directorate of Economics & Statistics, Rajasthan, Jaipur. p. 3.

13. Kalwar, (2008) : Wastelands : Planning for Development, Concept Publishing Company, New Delhi.

□□□

3

Population and Settlement

3.1 INTRODUCTION

An adequate water supply is generally recognised as an integral part for the health and welfare of the inhabitants of human settlements. The active and potential workers are known as human resources.[1] Human resources are the instruments of creating the capacity to create wealth.[2] In other words, man as the producer and consumer of natural resources, is the great resource by himself. "Natural environment in itself is a dynamic element but without human use it has no importance".[3] In this way, the activity of man is necessary to procure benefits from the unseen resources of nature. What is lacking is the recognition that water is one of the location factors in the urban and rural planning, a particularly relevant factor in a developing country like India, where financial limitations do not permit transportation of water from distant sources, especially in rural areas. So, it is an essential task to analyse the types and patterns of rural settlements and distribution of population for understanding the problem of water supply and its remedies. Rural settlement needs water mainly for domestic, livestock and irrigation but daily water demand for domestic and agricultural purposes varies according to the areas and seasons. It is also well known that the need for water increases proportionately as development of agricultural practices and population increases. Safe fresh water is needed for health and hygiene.

3.2 GROWTH OF POPULATION

The growth of population occurs either due to natural increase or due to net immigration. On village level, it is very difficult, rather impossible to trace the growth of population due to the lack of data. The first reliable and complete census was taken in 1891, but in the following pages population growth during the last century (1901-2001) only, has been discussed.

Table 3.1 : Growth of Population in Jaipur District

Year	Population	Absolute Variation	Percentage decadal variation	Absolute Variation over 1901	Percentage variation over 1901
1	2	3	4	5	6
1901	1212812	-	-	-	-
1911	1142522	-70290	-5.80	-70290	-5.80
1921	937601	-204921	-17.94	-275211	-22.69
1931	1045190	107589	11.47	-167622	-13.82
1941	1189737	144547	13.83	-23075	-1.90
1951	1524493	334756	28.14	311681	25.70
1961	1901756	377263	24.75	688944	56.81
1971	2482385	580629	30.53	1269573	104.68
1981	3436172	953787	38.42	2223360	183.32
1991	4722551	1286379	37.44	3509739	289.39
1991*	3887895	451723	13.15	2675083	220.57
2001	5252388	529837	11.22	4039576	333.08
2001	5252388	1364493	35.10	4039576	333.08

* without Dausa district (Dausa district separated in 1993 -1994)

Source : Directorate of Economics & Statistics, Rajasthan, Jaipur.

From the census report, it appears that the district population has shown a remarkable increment in its number for the last half century. The variation of total population has been positive. From the Table 3.1 it is clear that spectacular increase in numbers has not occured regularly. Mostly it increased during 1931 to 1991. The above Table show growth of population in Jaipur district, records three distinct periods of population variation : (1) from 1901 to 1931, (2) from 1931 to 1991 and (3) from 1991 to 2001. During the first period of 30 years, the decadal growth rate was very slow, the population decreased due to famine and epidemics. During second period (1931-91), the growth of population has been relatively rapid. After 1931 growth rate has been continuously increasing because of the availability and expansion of medical facilities and control over diseases etc. The increase being 289.39 percent in total population. During the third period (1991 to 2001) witnessed slightly slow growth, because in the 1993 Dausa district was separated from Jaipur district for the convenience of the administration, but over all growth of population has been relatively rapid in the period 1931 to 2001. The increase being 333.08 percent in the total population.

The period of 1901-1911 had a slight decrease of -5.80 percent in the total number of population. The period 1911-1921 also shows a slight decrease in the number of population. The decrease was -17.94 percent in total population during this decade. During 1921-31, the harvest on the whole was excellent and there were no serious epidemic outbreaks. These accounted for all around increase of 11.47 percent and the period of 1931-1941 also shows increase in the number of population. The increase was 13.83 in total population.

The decade of 1941-51 shows an uninterrupted increase in population. The percentage increase was 28.14 in total population. During 1951-61, the increase was 24.75 percent and in period of 1961-1971 increase was 30.53 percent in total population. After 1971, the 1971-91 decade maintained the trend of increase of population in total.

But the period of 1991-2001 had a slight increase of 11.22 percent in the total population because in the year 1993-1994 Dausa district was carved out from Jaipur district for the convenience of the government administration.

In general the increase in population was 333.08 percent in total population during last century (1901-2001), certainly a high increase against the limited resources of the area.

3.3 SPATIAL TEMPORAL VARIATION IN POPULATION GROWTH

Table 3.2 reveals that the maximum growth rate in Jaipur tehsil (+54.54) was followed by Bassi tehsil (+43.96). In Rajasthan's capital "Jaipur city" many unskilled, semi-skilled and skilled labourers came from the neighbouring districts in search of employment, while in Bassi tehsil the population has increased rapidly due to expansion of the capital city in western part. In the remaining tehsils the growth rate was below the district-average. The growth rate was the lowest in the tehsils of western and southern parts of the district viz. Sanganer, Phulera, Dudu and Phagi tehsils. These tehsils are agriculturally backward due to calcareous and saline soils. In Sanganer tehsil, average growth rate has decreased (-25.50) rapidly due to expansion of Jaipur tehsil.

Table 3.2 : Spatial Temporal Variation in Population Growth-2001

No.	Tehsil	Average Growth	Growth of Rural Population	Growth of Urban Population
1	Kotputli	+29.27	+29.65	+26.48
2	Viratnagar	+33.10	+33.43	+30.89
3	Shahpura	+25.79	+40.22	-27.00
4	Chomu	+31.07	+30.96	+31.65
5	Phulera	+25.23	+34.97	-1.42
6	Dudu	+20.32	+20.32	-
7	Phagi	+23.84	+23.84	-
8	Sanganer	-25.50	-14.69	-57.50
9	Jaipur	+54.54	-13.69	+59.37
10	Amber	+27.54	+27.54	-
11	J. Ramgarh	+31.12	+31.12	-
12	Bassi	+43.96	+43.96	-
13	Chaksu	+30.19	+28.17	+42.65
	Average	**+36.44**	**+25.80**	**+49.41**

Source : Directorate of Census Operations, Jaipur.

The pattern of rural population growth rate is almost the same as the aggregate growth rate. The maximum rural population growth rate was in Bassi (+43.96) followed by Shahpura tehsil (+40.22). While in Sanganer (-14.69) and Jaipur tehsil (-13.69), rural population has decreased rapidly due to expansion of Jaipur city.

The urban population growth rate (+49.41) is more than double in comparison with the rural population growth rate (+25.80) in the district. In urban areas, the population growth was the highest in Jaipur tehsil, followed by Chaksu and Chomu tehsils as in these towns basic amenities are growing at a faster pace.

3.4 DISTRIBUTION OF RURAL POPULATION

The distribution of population over space is governed by the general habitability of the area and its occupancy in historical perspective. The direct impact of physical environment explains the pattern of population spread[4]. Throughout history, the concentration of population was found on the water front points and fertile plains. The distribution of population in the study area follows the aforesaid base and reflects their clear impact on distribution of population. There is sharp difference in rural population distribution in two areas: high concentration of population in the northern upland tract and dispersed population towards the south plain tract.

3.4.1 The Northern Upland Tract

There is a homogeneous picture of rural population distribution in the northern upland tract. The great concentration of population lies in this region, which includes the Amber, Chomu, Kotputli tehsil in northeast Jamwa Ramgarh, Bassi tehsil and western quarter of the Phulera tehsil. It is one of the most salt productive regions of India and supports a dense rural population. The combination of fertile soil and water availability in the area, is suited for varied crops and its support to denser rural population.

Considering spatial perspective of rural population the upland can be divided into three groups distinguished by two levels of concentration-high and moderate.

3.4.1.1 The area of high concentration

The major concentration of population has been marked mainly around the urban centre and in the areas well connected with road and railway. Topography of water availability are the important factors in the concentration of population. These facts clearly indicate their effect.

The major concentration of population can be marked in the middle part of Amber tehsil. The main reason of concentration in this part is the location of Amber which is near to the Jaipur city. It provides the part-time job to the villagers in a circle of about 45 km, and also better agricultural facilities like seeds, agricultural implements, government aids and ready market.

Table 3.3 : Tehsil wise Rural and Urban Population in Jaipur District-2001

No		Rural Population			Urban Population			
	Tehsil	Male	Female	Total	Male	Female	Total	G.Total
1	Kotputli	156862	142772	299634	21354	18803	40157	339791
2	Viratnagar	62189	56235	118424	9050	8187	17237	135661
3	Shahpura	104150	93824	197974	14795	13375	28170	226144
4	Chomu	143480	132107	275587	26635	24082	50717	326304
5	Phulera	159193	148120	307313	42596	39395	81991	389304
6	Dudu	90441	84242	174683	-	-	-	174683
7	Phagi	84152	77396	161548	-	-	-	161548
8	Sanganer	68657	62622	131279	11571	10518	22089	153368
9	Jaipur	46412	42548	88960	1239711	1084608	2324319	2413279
10	Amber	137156	124750	261906	-	-	-	261906
11	J. Ramgarh	131017	119189	250206	-	-	-	250206
12	Bassi	120313	109656	229969	-	-	-	229969
13	Chaksu	84109	77005	161114	15253	13858	29111	190225
	Total	1388131	1270466	2658597	1380965	1212826	2593791	5252388

Source : Directorate of Census Operations, Jaipur.

In Phulera tehsil the whole western part comprising Jaipur, Kotputli, block have high concentration of population due to combination of better agricultural facilities and part time job in the largest salt lake town, Sambhar.

3.4.1.2 The area of moderate concentration

Besides high concentration rest of the rural areas of Bassi tehsil, Jamwa Ramgarh block and the western level plain area of the Chomu block show moderate concentration. All the settlements depend upon the local availability of water from the seasonal hill torrents and local rivulets and their presence has determined the incidence of rural population concentration of moderate sizes.

3.4.2 The South Plain Tract

Like other upland regions, due to inaccessibility and rough terrain which reports very sparse population, the southern-western part of Jaipur district is Chaksu, Phagi, Sanganer and Dudu block and all have similar characteristics. This part of district has sandy loam to loam texture on the surface while loam to clay meability is very low. This part of area has effected to fluorides and most of the area of this part is plain having poor drainage, so the life is not secure and easy there. These areas are restrictively suitable for human habitation.

3.5 IMPACT OF URBANISATION ON RURAL POPULATION

Urban growth and its impact on regional development is highly relevant today for developing countries like India, which are passing through various phases of economic development. Urbanisation influences the surrounding rural areas which can be seen in rural-urban fringe. Thus, urbanisation is equipped with potentialities of development of resources to solve in a better way the requirements of urban manufactured goods in the rural areas. In the context of regional development at the district level, attention has been paid to the various problems which are closely connected with the process of urbanisation and population growth.

Urbanisation naturally creates a strange behaviour and one would find that there is more impersonality, less mechanical consensus, more control on behaviour, less informality and high specialisation in urban areas.[5] There are two customary indices; to study the level and trends of urbanisation[6].

Impact of urbanisation in Jaipur tehsil is very low concentration of rural population (Table 3.3). It seems also to have been accelerated by growth in rural areas but not such as urban area, which necessitated either the sub-division of farms among several sons or the migration of surplus offspring to the cities. This movement to the large urban concentration got accelerated especially in the last century.[7] The rate of urbanisation in Asia has also been rapid in this century, but in many areas the increases have been from rather low base. Today more than 23.31 per cent of Indian population resides in cities.[8]

3.6 PATTERN OF RURAL SETTLEMENTS

Among the most visible and stable features, settlement has a distinct place especially in rural environment. It is a man made habitat on the earth's surface, representing an organized colony of human beings including the buildings in which they live or work or store or use them otherwise and the tracts and streets over which their movements take place[9]. The spatial characteristics of rural settlement are no doubt primarily determined by the physical environment and among all, water is a controlling factor on which all the human activity depends, especially when we focus our attention towards the site of the settlement. Although the structural growth is mostly influenced by the socio-economic achievements of the residents and their growing contact with the outer world in spatio-temporal perspective. In these contexts the rural settlement of the Jaipur district has been analysed. Being one of the most anciently settled regions of the world, the district is the true representative of the Jaipur tehsil.

3.6.1 Distributional Pattern of Rural Settlements

The distribution of villages can be better expressed with the village boundary tehsil-wise maps of the Jaipur district which clearly shows association pattern on the one hand, which is guided by the physical environmental factors such as relief, sources of water supply, drainage line, soil condition, etc. and on the other hand it is closely related to various socio-economic conditions like land use, land tenure, crop association, means of transport and communication etc. In general, ravines, *'user'* infested (infertile) lands, bad drainage and the danger of drought are the main restrictive forces for the uniform distribution of settlements over the plain; while the stony wastes, forest and difficult means of water supply have greatly restricted the choice of the settlement sites over the physical fragmented southern part of district. The water sources have played an important role in the selection of sites for human habitations. Water-front and dry sites are mostly characterised by large size compact villages like in Dudu, Phagi and Chaksu tehsil; while others like Jamwa Ramgarh and Bassi tehsil are regularly distributed along the water sources. The western part of the district is represented with uniform distribution along with preponderance of hamlets. In the southern and the eastern part, the situation has been slightly changed where villages become quite bigger and the distribution also becomes relatively regular. The clear-cut impact of terrain is visible in the north-eastern part of the district which is mostly forested. Patches of house-clusters are visible at the margin areas, thus giving rise to uneven and sometimes clustered distribution.

3.7 SEX RATIO

In Jaipur district females have remained less in comparison to males (897 females per 1000 males). The same pattern is found in rural areas (915 females per 1000 males) as well as in urban areas (878 females per 1000 males). Maximum male-female ratio in rural area is found in the western part of the district (Phulera 930, Chomu 921 and Phagi 920 females per 1000 males) while very low ratio is in Shahpura (901 females per 1000) tehsils.

Table 3.4 : Sex Ratio of Rural and Urban Population of Jaipur District 2001

No.	Tehsil	Female per thousand Male		
		Rural	Urban	Total
1	Kotputli	910	881	907
2	Viratnagar	904	905	904
3	Shahpura	901	904	901
4	Chomu	921	904	918
5	Phulera	930	925	929
6	Dudu	931	-	931
7	Phagi	920	-	920
8	Sanganer	912	909	912
9	Jaipur	917	875	876
10	Amber	910	-	910
11	J. Ramgarh	910	-	910
12	Bassi	911	-	911
13	Chaksu	916	909	914
	Average	915	878	897

Source : Directorate of Census Operations, Jaipur.

In urban area maximum male-female ratio is also found in the western part of the district (Phulera 925 females per 1000 males). The female population is very low in capital city of Jaipur (only 875 females per 1000 males) as males from surrounding areas and from other parts of the State come here for employment.

Besides, Jaipur is the centre of higher education, Industrials and many administrative offices of the State and Central Government are also situated. In tehsils of north-eastern part of the district male and female ratio is either medium or low.

3.8 LITERACY

Literacy leads to socio-cultural development of the people. Literate persons take part in various types of developmental activities, for example water supply programme is for common people but the method of implementation involves considerable technical inputs. Therefore it is a unique combination of degree of technological skill and welfare objective of the Government. In fact, there are 3 types of people involved in the water supply programmes. They are the technical people, semi-technical professionals and the non-technical hands. Broadly speaking, this would call for the incumbents not only from engineering side but even scientists, geographers, medical professionals, sociologists, chemists and communication experts. All these make the requirement of training and development complex and diverse in its contents. Therefore, the literate persons and human resource development activities in water sector acquires prominence.

In rural Jaipur population highest literacy is in Jaipur tehsil i.e. 40.89 percent. It is due to large rural population concentration in Jaipur city and the existence of a number of educational institutions. In the remaining tehsils literacy varies from 29.14 to 39.14 percent (Table 3.5).

The literacy rate in the district has improved over the last few decades but there is still a marked difference in the literacy rates of male and female. Female literacy rates are falling behind male literacy rates in all parts of the rural Jaipur district. The higher rates of population increase, with relatively higher literacy percentage among children and adults, require a new strategy to be evolved for implementation of the formal education system.[10]

Male literacy is 55.24 percent in rural Jaipur district. It is the highest in Jaipur tehsil (63.88 percent), followed by Chomu tehsil (59.76 percent). In the remaining tehsils, rural male literacy varies from 46.06 to 59.56 percent. The minimum rural male literacy is found in

Phagi tehsil (46.07 percent) and Chaksu tehsil (46.24 percent). It indicates that male literacy is less in south-eastern part of rural areas of Jaipur district.

Table 3.5 : Literacy Rates for Rural Population of Jaipur District by Sex

(Excluding Children in the Age Group 0-6) (2001)

	Tehsil	Persons	Males	Females
1	Kotputli	36.09	56.35	13.91
2	Viratnagar	34.36	53.91	12.45
3	Shahpura	37.92	59.56	13.89
4	Chomu	39.23	59.76	15.60
5	Phulera	36.39	58.39	12.62
6	Dudu	32.67	50.90	12.58
7	Phagi	29.46	46.06	10.94
8	Sanganer	38.46	59.44	14.63
9	Jaipur	40.89	63.88	15.66
10	Amber	35.39	57.15	11.14
11	J. Ramgarh	31.75	51.86	9.24
12	Bassi	33.47	54.76	9.72
13	Chaksu	29.14	46.24	7.99
Rural Jaipur District		**35.02**	**55.25**	**12.34**

Source : Directorate of Economics & Statistics, Rajasthan, Jaipur.

Female literacy is only 12.34 percent in rural Jaipur district. It is the highest in Jaipur tehsil (15.66 percent), followed by Phulera tehsil (15.60 percent). In the remaining tehsils female literacy varies from 7.99 to 14.63 percent. The minimum rural female literacy is found in Chaksu tehsil (7.99 percent), Bassi (9.72 percent) and Jamwa Ramgarh tehsil (9.24 percent). It indicates that female literacy is very less in rural areas.

3.9 DENSITY OF POPULATION

Land resources are limited and population is increasing rapidly. Thus it is necessary to assess the pressure of population on land. For this, various densities have been calculated and analysed.

3.9.1 Arithmetic Density

Arithmetic density is the ratio between the total population and the total geographical area. The arithmetic density of the district was 471 person per square kilometre in 2001. Tehsil wise analysis reveals that the arithmetic density varies from 131 to 4123 person per square kilometre (Table 3.6).

Table 3.6 : Arithmetic and Rural Density of Jaipur District (2001) (person per square kilometre)

No.	Tehsil	Arithmetic Density	Rural Density
1	Kotputli	407	368
2	Viratnagar	286	267
3	Shahpura	475	429
4	Chomu	474	414
5	Phulera	267	227
6	Dudu	131	131
7	Phagi	143	143
8	Sanganer	244	225
9	Jaipur	4123	231
10	Amber	310	310
11	J. Ramgarh	247	247
12	Bassi	356	356
13	Chaksu	244	210

Source : Directorate of Economics & Statistics, Rajasthan, Jaipur.

The very high density (400-500) is concentrated approximately in the central and north-western parts of the district as it includes the capital city Jaipur, Chomu, Shahpura and Kotputli tehsil. This is scattered in two parts viz. north-western and central parts of the district. The northern part is occupied by the basins of the Sota & Sabi rivers having fertile, alluvial deep and non-calcareous soils with high intensity of irrigation. Medium high arithmetic density i.e. 300-400 persons per square kilometre corresponds to medium to high irrigation intensity in Amber and Bassi tehsil. The large area of Jaipur district is medium-low arithmetic density (200-300) which is concentrated in Phulera, Sanganer, Chaksu, Jamwa Ramgarh and Viratnagar tehsils. Low arithmetic density (100-200) is concentrated only in the south-western part of the district.

This part of the district has sandy loam to loam texture on the surface with loam to clay loam and deep calcareous sub soil, PH reaction is strong and permeability is very low. 'Kankars' are present below sixty centimetres depth. Most of the area of this part is plain having poor drainage.

3.9.2 Rural Density

Rural density is the ratio between rural population and rural areas. The pattern of rural density is almost same as that of the arithmetic density except central and northern parts of the district. In the central part the urban population is more concentrated due to the location of the capital city. Five tehsils viz. Jamwa Ramgarh, Dudu, Phagi Bassi and Amber have no urban population, therefore, in these tehsils rural and arithmetic densities are exactly the same. In the western part Sambhar lake and Phulera towns and in the north western part Chomu and Shahpura towns are situated, so the rural density is slightly less than the arithmetic density (Table 3.6). In the northern part, Kotputli town is situated, so rural density is less than the arithmetic density.

3.9.3 Agricultural Density

Agricultural density is the ratio between agricultural population and agricultural land i.e. *Net Sown Area* (NSA). Very high agricultural density (more than 130) is concentrated in the north-eastern part i.e. Jamwa Ramgarh tehsil. This tehsil has the lowest *Net Sown Area* in the district due to the Aravalli hills. High agricultural density (100-130) is scattered in north-east part of the district. This part also has low *Net Sown Area* due to gully eroded land which runs along the eastern boundary.

In western part of the district, the agricultural density is low to very low. This part has more *Net Sown Area* due to peneplain area. Remaining tehsils have medium agricultural density. These tehsils lie in the north-eastern part of the district.

3.9.4 Physiological Density

The physiological density is a ratio between population and cultivated land. This density is more important than other densities of population as this density represents the population pressure on cultivated land. The northern part has more physiological density while the southern part has low density. The medium physiological density (250-300) is concentrated in the central part. In southern part of the district except south-eastern part the soils are calcareous and saline, rain-fall and irrigation intensity are also the lowest in the district.

3.10 OCCUPATIONAL STRUCTURE

The occupational structure is the true index of the nature of activities practised in the district. According to the census of India, the main workers i.e. those who have worked for six months or more in the year, preceding the enumeration, have been further categorised as cultivators, agricultural labourers and household industry workers as other workers[11].

3.10.1 Cultivators

A person is a cultivator if he is engaged in the cultivation of land. Cultivation involves ploughing, sowing and harvesting and covers production of cereals, millets, pulses, oil-seeds, fibre crops (Jute, Cotton) spices, sugarcane etc. but does not include fruit growing or plantation like tea, coffee and rubber.

In the year 2001, 66.05 percent female and 34.5 percent male population were engaged as cultivators in Jaipur district. More than 70 per cent cultivators of the working population are found in the eastern part of the district. It is due to the fertile land with high irrigation intensity. The lowest percentage of cultivators is found in Jaipur tehsil i.e. only 4.19 per cent male and 15.95 percent female of the total working population.

Rajasthan's capital, Jaipur city, provides job opportunities to rural population of the Jaipur tehsil and to its adjoining tehsils in secondary and tertiary activities. Therefore, the percentage of cultivator ranges from 4.19 to 65.61 percent male and 15.95 to 85.15 percent female.

In rural area this percentage ranges from 53.89 to 72.05 percent male and 70.71 to 86.71 percent female and in urban area this ranges from 0.77 to 36.23 percent male and 2.68 to 71.43 percent female (Table 3.7).

Table 3.7 : Population of Cultivators, Agricultural, Household and Other worker of Jaipur District[12] (in percentage) 2001

Jaipur District		Cultivators Male	Cultivators Female	Agricultural Male	Agricultural Female	Household Male	Household Female	Other Male	Other Female
	Rural	61.73	80.9	8.13	11.71	2.98	2	27.16	5.39
	Urbon	3.75	14.27	1.07	2.3	4.11	8.62	91.06	77.01
	Total	34.5	66.05	4.82	9.61	3.51	2.99	57.17	21.35
Tehsil		**Male**	**Female**	**Male**	**Female**	**Male**	**Female**	**Male**	**Female**
1 Kotputli	Rural	65.81	80.36	8.81	14.96	2.32	1.01	23.06	3.67
	Urbon	16.97	39.96	7.65	13.26	3.59	14.39	71.79	32.39
	Total	59.88	78.26	8.67	14.87	2.48	1.71	28.98	5.17
2 Viratnagar	Rural	68.00	80.56	10.51	13.65	2.87	1.63	18.63	4.16
	Urbon	36.23	69.87	16.60	21.89	4.71	1.52	42.46	6.73
	Total	64.24	79.24	11.23	14.66	3.08	1.62	21.45	4.48
3 Shahpura	Rural	58.52	77.82	6.73	10.83	4.80	4.91	29.94	6.44
	Urbon	30.90	61.33	1.88	8.59	9.16	12.99	58.06	17.09
	Total	52.66	75.40	5.70	10.50	5.73	6.10	35.91	8.00
4 Chomu	Rural	61.13	86.47	5.89	6.02	3.55	3.15	29.43	4.36
	Urbon	19.55	71.43	5.54	6.10	7.05	6.67	67.86	15.81
	Total	54.76	84.89	5.84	6.03	4.09	3.52	35.31	5.56
5 Phulera	Rural	62.05	82.64	8.62	11.75	3.25	1.40	26.08	4.21
	Urbon	12.68	30.88	4.79	8.26	4.54	9.84	78.00	51.02
	Total	49.25	76.25	7.63	11.32	3.58	2.45	39.53	10.00
6 Dudu	Rural	60.08	75.58	8.80	16.25	2.71	0.96	28.41	7.21
	Urbon	-	-	-	-	-	-	-	-
	Total	60.08	75.58	8.80	16.25	2.71	0.96	28.41	7.21
7 Phagi	Rural	63.23	70.71	10.47	21.03	2.31	1.34	23.99	6.92
	Urbon	-	-	-	-	-	-	-	-
	Total	63.23	70.71	10.47	21.03	2.31	1.34	23.99	6.92
8 Sanganer	Rural	53.89	76.80	9.07	12.63	1.91	1.26	35.12	9.31
	Urbon	15.54	53.52	2.22	2.79	6.19	16.64	76.05	27.05
	Total	44.04	73.25	7.31	11.12	3.01	3.61	45.63	12.02
9 Jaipur	Rural	56.06	81.41	7.25	9.59	2.73	2.53	33.96	6.47
	Urbon	0.77	2.68	0.35	0.72	3.74	4.84	95.14	91.76
	Total	4.19	15.95	0.78	2.22	3.68	4.45	91.35	77.37

Contd...

...Contd.

Tehsil		Male	Female	Male	Female	Male	Female	Male	Female
10 Amber	Rural	59.66	85.15	6.89	8.90	3.07	2.10	30.39	3.85
	Urbon	-	-	-	-	-	-	-	-
	Total	59.66	85.15	6.89	8.90	3.07	2.10	30.39	3.85
11 Ramgarah	Rural	61.47	77.95	9.22	12.87	3.73	3.09	25.59	6.09
	Urbon	-	-	-	-	-	-	-	-
	Total	61.47	77.95	9.22	12.87	3.73	3.09	25.59	6.09
12 Bassi	Rural	61.41	84.23	7.69	8.73	3.02	2.58	27.87	4.46
	Urbon	-	-	-	-	-	-	-	-
	Total	61.41	84.23	7.69	8.73	3.02	2.58	27.87	4.46
13 Chaksu	Rural	72.05	85.11	6.49	9.29	2.24	1.17	19.22	4.42
	Urbon	22.52	60.56	2.56	8.26	6.86	7.89	68.06	23.29
	Total	65.61	83.78	5.98	9.24	2.84	1.54	25.57	5.44

Source : Directorate of Economics & Statistics, Rajasthan, Jaipur.

3.10.2 Agricultural Labourers

A person who works in another person's land for wage in money or share is categorised as an agricultural labourer. In Jaipur district only 4.82 percent male and 9.61 percent female population is working as agricultural labourers. The maximum concentration of agricultural labourers is in the western and southern parts of the district particularly in Phulera, Dudu, Phagi and Bassi. All these tehsils are agriculturally and industrially backward, so the percentage of agricultural labourers is more. In Jaipur and its surrounding tehsils, the percentage of agricultural labourers is less. Therefore, the percentage of agricultural labourers ranges from 0.78 to 11.23 percent male and 2.22 to 21.03 percent female. In rural areas this percentage ranges from 5.89 to 10.51 percent male and 6.02 to 21.03 percent female and in urban area this ranges from 0.35 to 16.60 percent male and 0.72 to 21.89 percent female (Table 3.7).

3.10.3 Household Industry Workers

Household industry is defined as an industry conducted by the head of the household himself or by the members of the household at home or within the village in rural areas and only within the precincts of the house where the household lives in urban areas. The main criterion of the household industry is the participation of one or more member of a household. A household industry relates to production, processing, servicing or making and selling of goods.

Only 3.51 percent male and 2.99 percent female population were engaged in household industry in 2001. The maximum percentage of population engaged in household industry was in Shahpura tehsil, followed by Chomu, Jaipur and Sanganer tehsil, because in these areas so many small scale industries have emerged for making consumable home products, plastics goods, baskets, ropes and other. In Jaipur many families do their work such as jewellery etc, in their houses. Therefore, the percentage of household industry workers ranges from 2.31 to 5.73 percent male and 0.96 to 6.10 percent female. In rural area this percentage range from 1.91 to 4.80 percent male and 0.96 to 4.91 percent female and in urban area this range from 3.59 to 9.16 percent male and 1.52 to 16.64 percent female (Table 3.7).

3.10.4 Other Workers

All workers excluding cultivators, agricultural labourers and those engaged in household industry are classified as "*Other workers*". This category includes factory workers, plantation worker, those in trade, commerce, business, transport, mining, construction, political or social workers, all government servants, municipal employees, teachers, priests, entertainment artists, etc. Workers of this category can be either main workers or marginal workers. Marginal workers are those who have worked at any time but the period for which they had worked is less than 183 days.

In Jaipur district more than fifty percent working population is categorised as other workers. This is the only district in the whole State where the percentage of other workers is very high. The maximum concentration of marginal workers is in Jaipur tehsil (91.35 percent male and 77.37 percent female) followed by Sanganer and Phulera tehsils. In Jaipur city, many Central, State and quasi-government offices are situated. Besides, most of the workers are engaged in trade, business, construction, industries etc. In Phulera tehsil, the largest salt lake of the country is situated near Sambhar lake town. This salt lake belongs to Hindustan Salt Limited and about two thousand employees work over here. Phulera is the main railway junction which lies on the Delhi-Ahmedabad, Jaipur-Jodhpur broad gauge railway lines and Delhi-Udaipur meter gauge railway line and is only 60 kilometres away from Jaipur city. Daily 1000-1500 persons come to Jaipur city from Jamwa Ramgarh, Bassi, Sanganer and Amber to do the work of "Building Construction". Sanganer tehsil is in the south of Jaipur city and Sanganer town is only 11 km away from the city. Therefore, many persons come from this tehsil for construction work and as casual labourers etc. Amber tehsil is also an adjoining tehsil of Jaipur tehsil and it is only a few kilometres away from Jaipur city. In remaining tehsils the percentage *of* other workers is less than one fourth of the total working population.

3.11 STANDARD OF LIVING

An all-inclusive step towards the transformation of the economy was the initiation of development plans with the launching of First Five Year Plan for the country in April 1951. The various measures of land reforms and policy of providing support to agricultural prices, aimed at ensuring full returns of labour to the agriculturist and thus raising the standard of the large majority of population, is dependent on agriculture. The introduction of Community Development Blocks and the Scheme of Democratic Decentralisation, made the villagers partners in the efforts initiated by the government for upliftment of the rural areas. The incentives provided to the cottage, village and small industries helped to diversify the means of livelihood both for rural and urban population, thereby adding to their incomes and making lives richer and more varied. The extension of community services in the form of more schools, dispensaries and hospitals as well as the increasing means of transportation and communication have been going a Iong way in increasing the living standards of the residents of the district and the State. People eat better food, live in better-built houses, and enjoy other facilities provided by the government and also have increased income. The common villagers, to a great extent, have to get out of the clutches of the evil of debt and the signs of their better material conditions are specially visible on ceremonial occasions. The age-long disabilities of the down-trodden sections of the Scheduled Castes and Scheduled Tribes are gradually being removed with the help of the special schemes taken up by the "Social Welfare" and other departments of the State Government.

REFERENCES

1. Mandal, R. B. (1976) : *Nature of Population Geography, Dimension in Geography.* Associated Book Agency, Patna, p.96.

2. Bhindouriy, B.P.S. & Dua, S.C. (1986) : *Rural Development Strategy and Perspective.* Annual Publications, Delhi, India, p.53.
3. Ahmed, K.S. (1946) : "Environment and the Institution of Population in India", *Indiana Geographical Journal,* Vol. VI, p. 3[illegible].
4. Singh, R.B. (1975) : "Rajput Clan-Settlement in Varanasi District", *National Geography Society India, Research Publication* No.12, p.15.
5. Mullick, U.C. (1979) : *Development of Small Town for Planned Ubanisation and Economic Growth : Place of Small Towns in India.* Inter-India Publication, Delhi, p.55.
6. Chandrashekhar, C.S. et. al. (1972) : *Growth Force in Regional Strategy.* University of Maysor, p.55.
7. Ehorlich, R. Paul (1970) : *Population Resources Environment-Issues in Human Ecology* . W.M. Freeman and Co., San Francisco, p.38
8. Sinha, U.P. (1985) : *Planned Development of Resources in a Developing Region,* Inter India Publication, New Delhi, p.38
9. Singh, R.L. (1962) : "Meaning Objective and Scope of Settlement Geogrphy", *National Geography Journal,* Vol. VII, p.12.
10. Dhabriya, S.S. & Saini, R.R. (1986) : *Existing Land Use System and Possible Strategy to Evolve an Optimum Land Use Plan for Jaipur District.* Social Policy Research Institute, Jaipur, p. 15.
11. Census of India (1991) : *Jaipur District Census Hand Book.*
12. Directorate of Economics & Statistics (2001) : *District Statistical Abstract-2001,* Directorate of Economics & Statistics Rajasthan, Jaipur, p. 20.

4

Water Resource Potentiality and its Utilisation Pattern

The main objective of water resource planning is to make the most effective use of the available water resources to meet all the foreseeable short and long-term needs of the area[1]. The water requirement for the growing human and animal population, irrigation and industrial use is increasing at an alarming rate. The extent to which water supply can be increased depends on the potentiality of water resources of the area[2]. Rainfall is an essential factor for water supply. The most appropriate geographical unit for water resource planning are mainly grouped under two heads : (a) Surface water supplies mainly including running water from springs, streams and rivers; stationary water from lake and ponds or from reservoirs and tanks, (b) underground water supplies. If the rainfall is sufficient, the supply remains normal, but failure of rainfall generally leads to water-difficulties especially if the sources of supply are small in capacity. The consideration of geohydrological condition of the area is a pre-requisite before assessing the sources of water because it governs the pattern of the efficiency of water tapability and recharge condition.

4.1 SURFACE WATER SOURCES

The district has good surface water potentiality. About 80 to 90 percent surface run-off occurs in the hill areas, and maximum infiltration as ground water occurs in piedmont and plains where the slope of ground is less. Only small fraction of the rainfall becomes ground water and rest of the water is lost through evapotranspiration, run-off and soil moisture, and remaining goes into tanks and dams.

Surface water is available in the river and streams, natural and artificial lakes. Using remote senseing data products, the following hydrological units have been identified and mapped in Jaipur district (Figure 4.1).

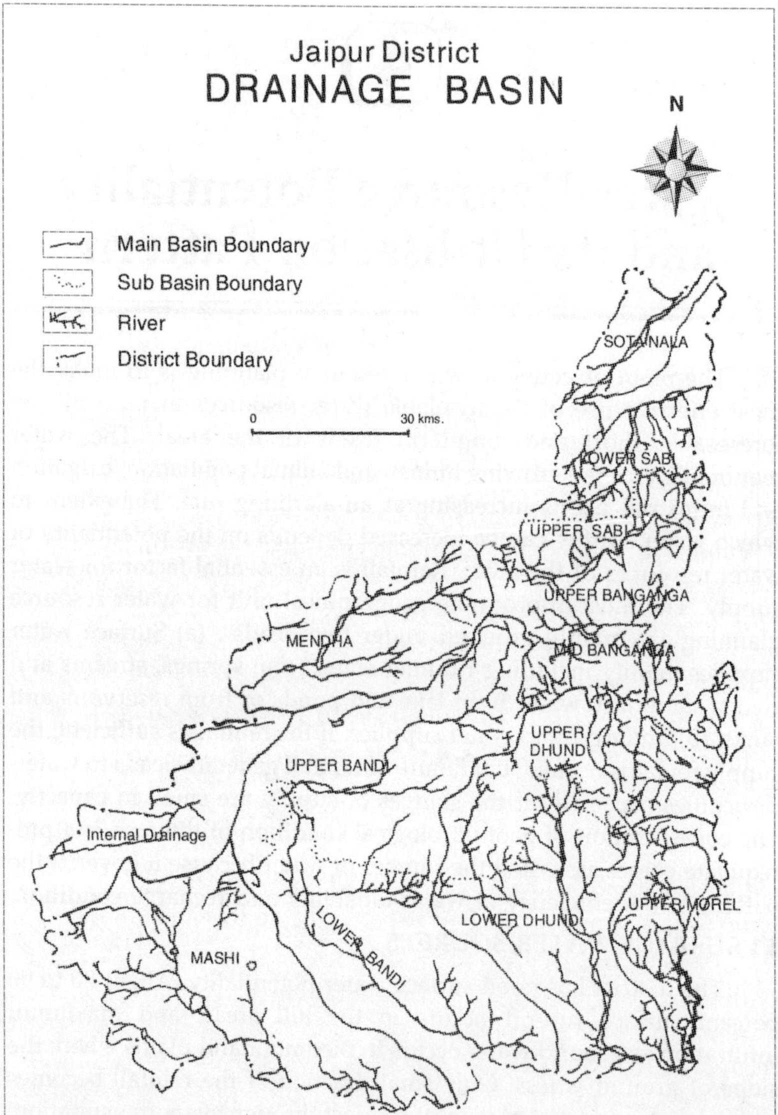

Figure : 4.1

The studies of the various surface sources cover five river basins, namely : Mendha River Basin, Sabi River Basin, Banganga River Basin, Mashi River Basin and Morel River Basin.

Mashi basin and Morel basin are sub-basins of river Banas, which passes through districts of Tonk and Sawai Madhopur. Mashi river joins river Banas near Tonk and Morel river joins river Banas near Olwara in Sawai Madhopur district. Mendha river terminates into Sambhar lake near Phulera in Jaipur district where salt manufacturing unit has been installed. Sabi and Banganga basins cover the northern and north eastern parts of Jaipur district. The Table No. 4.1 shows the basin wise runoff and surplus water availability of these five river basins.

4.1.1 Mendha River Basin

The Mendha and its tributaries flow in Jaipur district and traverse Amber and Phulera tehsil after originating from the western slopes of Uaipurwati range near village Deorala in Danta Ramgarh tehsil of Sikar district. It flows in Southwest direction and terminates in Sambhar salt lake. The length of this river is 110.80 km. The total catchment area of the sub-basin is 4185.60 sq.km (Table 4.1). The basin is situated in Ajmer, Jaipur, Sikar and Nagaur districts.

The catchment area is mostly plain. The annual average monsoon rainfall of the basin is 398.78 mm, the estimated run off from 1635 sq.miles of the catchment area works out to be 2753 Mcft or 76.60 Mcum i.e. the runoff yield from the catchment area is at the rate of 1.7 Mcft/Sq.mile. Because of the low yield from the catchments and the area being plain, the basin is termed as 'Bad'.

As against estimated runoff of 76.60 Mcum, existing tanks utilise 24.69 Mcum while expected utilisation through ongoing schemes is 6.51 Mcum and 1.55 Mcum by Panchayat Samiti tanks. Thus total utilisation comes to 32.75 Mcum leaving a balance of 43.85 Mcum to join Sambhar Lake. The entire water potential of the sub-basin goes into Sambhar lake and no attempt has been made to utilise it except at Jaitpura and Kaladera, as its utilisation will effect the output of Sambhar Lake. Hence it is concluded that there is no possibility in future also to utilise its water potential. Table No 4.1 shows the salient features of Mendha River Basin.

4.1.2 Sabi River Basin

Sabi river originates from Neem Ka Thana tehsil of Sikar district and passes through Viratnagar and Kotputli tehsil of Jaipur district. One of the Tributaries of Sabi river rises from the hills of Viratnagar village Barijori in district Jaipur and terminates at Dausa Bund in Haryana State. The total length of the river is 222 km. The river flows for a total length of 157 km. in Rajasthan. Its major tributary in Rajasthan, joining on its left flank, is river Sota.

The total catchment area of this basin is 4813 sq.km. in Rajasthan, (Table 4.1) out of which an area of 3842.56 sq.km. is intercepted by a number of existing, ongoing and proposed schemes. The remaining 970.44 sq.km. of catchment area if tapped will effect the storage capacity of a dam constructed on this river at Bewadi in Haryana State. The basin is situated in Jaipur, Sikar and Alwar districts.

The annual average monsoon rainfall is 579.62 mm., the estimated runoff from 4813 sq.km. catchment area works out to be 7551.78 Mcft or 210.12 Mcum i.e., the run off yield from the catchment is at the rate of 1.6 Mcft/sq.km. But if we consider the head reach of the catchment i.e. 828.8 sq.km. the rainfall yield works out to be satisfactory. Hence, the catchment is termed as average in upper reaches and 'Bad' in lower reaches.

Out of the total estimated runoff of 210.12 Mcum, utilisation by existing tanks is 89.40 Mcum, 26.07 Mcum by ongoing schemes, 73.65 Mcum by proposed schemes and 4.2 Mcum by Panchayat Samiti tanks, leaving only 16.80 Mcum as net balance. Hence it is concluded that there is no positive indication to intercept the balance water for drinking purposes as it effects the storage of dam on down stream side of the river in Haryana.

4.1.3 Banganga River Basin

Banganga river originates from Aravalli hills near Viratnagar tehsil, of Jaipur district, then flows towards south up to village Ghat, then towards east through partly hilly and partly plain terrain. The basin area of the river covers the parts of Jaipur, Alwar, Sawai Madhopur, Dausa and Bharatpur districts. The total length of this river is 240 km. Important tributaries to this river are Gumthi Nallah and Suri river.

Table 4 .1 : Estimation Of Surface Water Sources in Jaipur District (1997-1998)

No.	Name of Basin/ Sub-Basin	Catchment Area (Sq.Km.)	Annual Average Rain Fall (mm)	Type of Catchment	Total Available Run off (Mcum)	Surface Water Draft For Irrigation Purpose (Mcum)	Net Surface Water Balance (Mcum)	Remark
1	Mendha River Sub - Basin	4185.60	398.78	Bad	76.60	32.75	43.85	Water potential is not utilised completely.
2	Sabi River Basin	4183.00	579.62	Bad & Average	210.12	193.32	16.80	Only 80% of total catchment is tapped.
3	Banganga River Basin	6668.80	556.26	Good & Average	440.46	455.00	-	Water potential is completely utilised in Ramgarh Bund.
4	Mashi River Sub - Basin	6919.68	539.75	Bad	304.60	216.20	88.40	Balance water will be utilised in proposed Isreda project.
5	Morel Sub - Basin	5921.00	606.04	Average	517.42	329.77	187.65	Balance water will join river Banas and will contribute to the storage for the proposed Hydel project on Chambal.
	Total				1549.20	1227.04	336.70	

Source : Irrigation Department, Govt. of Rajasthan, Jaipur.

Total catchment area of the basin is 6668.8 sq.km. (Table 4.1) A total area of 6423.04 sq.km. is intercepted by a number of existing tanks in which Ramgarh Bund being the major one, meets the drinking water requirement of a major part of Jaipur.

The annual average monsoon rainfall is 556.26 mm. The total available runoff from the catchment area is 440.46 Mcum, i.e. rainfall yield of the basin is 2.4 Mcfl/sq.km. Catchment area is hilly in its upper reaches with an yield of 9 Mcft/sq.km. In the intermediate reaches it is partly hilly and partly plain resulting in a yield of 6 Mcfl/sq.km. and in its lower reaches being plain, the yield works out to be 4 Mcft/sq.km. Hence, the catchment area is termed as "Good" in its upper reaches and "Average" in its lower reaches.

As against the available runoff of 440.46 Mcum, a total of 373.68 Mcum is utilised by existing tanks. The ongoing and proposed schemes are expected to utilise 16.44 Mcum and 56.15 Mcum respectively leaving no balance of water from the basin.

This basin is already tapped by construction of Ramgarh Bund and under previous project wherein the height of the dam was increased to store more water. But the performance of this Reservoir has not been satisfactory. In last 18 years there had been very poor river inflow in the Ramgarh reservoir. Thus the production had been reduced considerably with an average withdrawal of 27 MLD from Ramgarh Bund system. It is also noted that during the year 1990 to 1997 it was not possible to draw more than 1.30 MLD and after the year 1997 practically nil from Ramgarh lake. Besides this, there are about 143 existing tanks in the basin and as such the yield of the basin stands fully utilised.

4.1.4 Mashi River Basin

The river Mashi which comes under sub-basins of river Banas originates from the hills near Kishangarh in Ajmer district and ultimately joins river Banas near Tonk district. The total length of the Mashi river is about 75 km. The sub-basin is situated in Ajmer, Jaipur and Tonk districts. The main tributaries of Mashi river are Dudu and Bandi.

The total catchment area of this sub-basin is 6919.68 sq.km. Out of which an area of 5298 sq.km. is intercepted by number of existing tanks. The catchment area in upper reaches is partly hilly and partly plain but in its lower reaches it is plain.

The annual average monsoon rainfall is 539.75 mm. (Table 4.1) The estimated run off from 6919.68 sq.km. of bed catchment for a rain fall of 539.75 mm works out to be 304.6 Mcum. Therefore, rainfall yields per sq.km. works out to be 1.7 Mcft.

As against the available run off of 304.6 Mcum, a total of 207.56 Mcum has been utilised by the existing tanks, 6.11 Mcum has been utilised by Panchayat Samiti tanks and 2.53 Mcum by the ongoing schemes, leaving a balance of 88.4 Mcum to join river Banas. The balance water can be utilised in proposed Isarda Dam to meet the future water requirements of Jaipur district or by proposed conventional schemes in the basin within the Jaipur district before it enters into district Tonk. Hence this river basin forms an effective and useful source of water to be given proper attention to utilise it to its full potential in future.

4.1.5 Morel River Basin

The river Morel originates from the hills near Dharla and Chainpura village in Bassi tehsil of Jaipur district. For a length of 64 km this river is named as Dhund. At a length of 83 km from Sooli village in Lalsot tehsil of Dausa district it joins the river Banas, and has been named as 'Morel'. The important tributaries of the river are Dhund and Karal. Morel sub-basin is situated in Jaipur, Dausa, Tonk and Sawai Madhopur districts.

The total catchment area of the sub-basin is 5921 sq.km. Out of which, an area of 4092.87 sq.km. is intercepted by a number of existing tanks.

Annual average monsoon rainfall is 606.04 mm. The estimated runoff from 5921 sq.km. of catchment area is classified as "Average" and works out to 517.42 Mcum.

As against the total available run off of 517.42 Mcum, a total of 329.77 Mcum has been utilised by existing tanks, ongoing and proposed schemes and Panchayat Samiti tanks, leaving a balance of 187.65 Mcum which contributes to the storage of proposed Hydel project at Rahughat site on Chambal. Alternatively the surplus water can be made available for drinking purposes by constructing a dam across the river before it joins the Banas river.

4.2 DEPOSITS IN RIVERS BASIN

Almost the entire Jaipur district, except for the area in the south-western and southern parts is underlain by quaternary sediments of aeolian origin with thickness varying from 10 to 100 metres, except for small outcrops of quartzite. The alluvial fills are found as narrow strips along the present river courses and in their abandoned channels.

In the greater part of the river courses, the alluvium consists of reworked aeolian sand. Gravels, pebbles and cobbles in the alluvium are confined to those sections of the river courses where the quartzite hills form the major part of the catchment. Gravely deposition occurs also as reported in Sotanala, a tributary of Sabi river.

The alluvium in the area consists of sand, silt, clay, kankar and gravel. Kankar in general is found throughout the alluvial thickness. Gravels in the area are generally comprised of kankar, quartz and ferromagnesium material.

The blown sand is represented by stabilised and drifting sand dunes. The aeolian sand rests on the recent alluvium near river courses (generally on the eastern banks) or on the older consolidated rocks. The blown sand is fine to very fine grained with silt and little clay. Being loose, unconsolidated and uniform in texture, it has high porosity and permeability. It readily absorbs water in the pore spaces.

Blown sand is found in the entire district but is more prominent in the western part of the district comprising the area towards east of Sambhar town and in the eastern part of the district comprising areas of Chaksu.

4.3 SURFACE WATER BALANCE

The Jaipur district is drained by five river basins as explained earlier. From the Table No. 4.1, it is seen that the total available run off from these five river basins has been estimated as 1549.20 Mcum, out of this a quantity of 1227.04 Mcum is utilised for irrigation. Therefore, only 336.70 Mcum of surplus water is available for further utilisation.

As the surplus water available from Mendha river basin is presently being utilised for salt production in Sambhar lake, so there is no possibility of utilising this water for drinking purposes. In case of Sabi river basin, the available surplus water is not in significant quantity, so that it can be utilised for drinking purposes. It is always necessary to maintain minimum flows in the river, so that the ground water get recharged. Hence it is concluded that the tapping of Sabi river basin is of no use.

From the table 4.1 it is seen that in Mashi and Morel river sub-basins enough surplus water is available to utilise it further for drinking purposes. Some proposals have been worked out to store surplus water available from Mashi river sub-basin in Isarda Dam project. In case of Morel river sub-basin a large quantity of surplus water i.e. 336.70 Mcum is going waste and some proposals has been made by State Irrigation Department to utilise this water in proposed Hydel project at Chambal. But in view of the future population growth rate and future water demand for drinking purposes, it is recommended that the surplus water available from Morel river basin should be utilised to meet the drinking water needs of Jaipur district by constructing a dam across the river before it joins Banas river.

4.4 HYDROLOGICAL CYCLE

The hydrological cycle is by no means a simple link but a group of numerous areas which represent the different paths through which water in nature circulate and is transformed.[3] As water changes its state and moves from place to place, a number of processes and energy exchanges occur. These processes and changes of state are going on all the time, however, the intensity varies. The hydrological cycle undergoes various complicated processes of evaporation, precipitation, interception, transpiration, infiltration, percolation, storage and run-off.[4]

4.4.1 Evaporation

Evaporation is the process by which water is changed from the liquid into the gaseous state through the transfer of heat energy.[5]

Hydrological data reveals that maximum evaporation occurs in the months of May (25.74 cm) and June (24.26 cm) and minimum in December (6.78 cm) in Jaipur district (Figure 4.2). The annual evaporation is more than 174.55 cm which is three times of the average annual rainfall. Less forest cover means higher evaporation losses. Whatever run-off is recorded in the different rivers and on the surface, is a result of the precipitation in the form of rainfall. Net run-off is a resultant of rainfall, evaporation losses, seepage, etc. The figure 4.2 highlights the hydrological conditions taking into account of the rainfall and evaporation only. It is evident from the table that whatever run off is recorded in the different river systems is mostly during the months of July and August, the other months are the months of deficit flow.

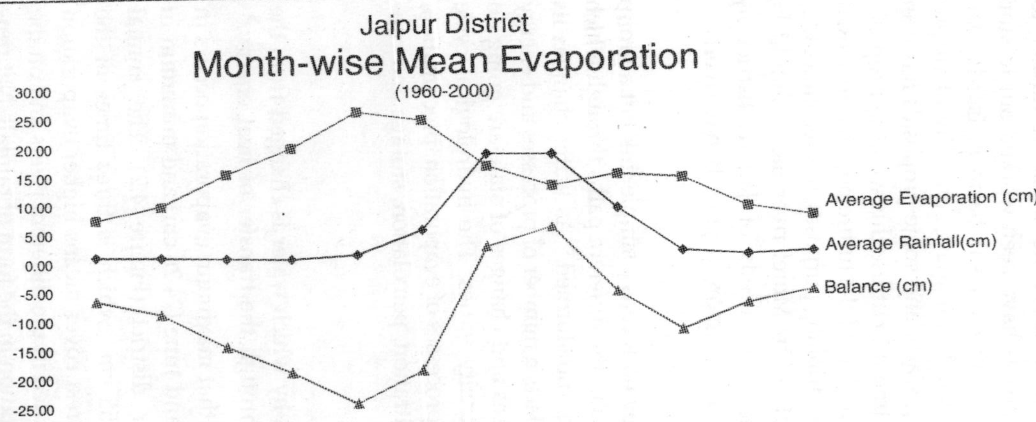

	Jan	Feb	Mar	Apr	May	Jun	Jul	Aug	Sep	Oct	Nov	Dec
Average Rainfall(cm)	1.12	0.90	0.59	0.36	0.99	5.13	18.27	18.07	8.52	0.99	0.19	0.57
Evaporation (cm)	7.56	9.82	15.27	19.64	25.74	24.26	16.17	12.64	14.48	13.72	8.47	6.78
Balance (cm)	-6.44	-8.92	-14.68	-19.28	-24.75	-19.13	2.10	5.43	-5.96	-12.73	-8.28	-6.21

◆ Average Rainfall(cm) ▦ Evaporation (cm) ▲ Balance (cm)

Evaporation Equation based on Dalton's Law.[6]

E = C (ew - ea)

Where :

C = Coefficient depending on various uncounted factors affecting evaporation of

Figure 4.2

4.4.2 Run-off

The quantity of (i.e. the portion of rainfall) water running in the streams is called run-off.[6] The run-off in a stream depends upon such factors as topography, soil, geological structure, temperature, character and intensity of rainfall. More run-off occurs in the hilly region while maximum infiltration of ground water takes place in piedmont zone and plain areas. The run-off is recorded in the Banganga catchment i.e. 440.46 Mcm, Mashi (304.60 Mcm.), Sabi (210.12 Mcm.), Mendha (76.60 Mcm.) and Morel (517.42 Mcm.). For the development of the water resources in the district harnessing of this run-off is quite necessary, by making check dams on the rivers. Check dams should be constructed of sizeable quantity so that river flow may be monitored and the whole catchment areas gets benefited.

4.5 SUBSURFACE GEOLOGY

Thirty two number of exploratory tube-wells have been constructed (and four slim holes are drilled so far) in Jaipur district by Central Ground Water Board for ground water resource evaluation in Banganga and Morel river basins. Apart from these tube-wells PHED and other agencies have constructed a number of tube-wells in the district. The sub-surface geology of the district was studied based on the results of exploratory tube-wells and other data available. The areas towards west of northeast-southwest trending ridges, are characterised by alluvial thickness of more than 50 m. (Figure 4.3A). The alluvial thickness seems to be gradually increasing from west to east attaining maximum thickness near wind ward side of ridges. In the area covering south western and southern part of the district, the alluvial thickness is less than 10 m and in this area comprising Naraina, Sakhun, Dudu, Mozmabad, Phagi, Chaksu, the granitic gneiss is encountered in dugwells. The maximum thickness of alluvium encountered (87.7 m) was in the exploratory well drilled.

In northern part of the district altitude of bedrock (Figure 4.3B) ranges from 325 mamsl to 425 mamsl indicating that bedrock slopes from southern to northern direction. In north-western and western part of the district altitude of bed rock ranges from 400 to 300 mamsl and though the bedrock gradually slopes in the south westerly direction there is a depression in the bedrock around Phulera and Jairampura where the altitude of bedrock ranges between 325 mamsl to 400 mamsl.

4.6 Ground Water

Ground water is the main source of irrigation and drinking water supply in the district. Ground water differs from sub-surface water such as capillary water and soil water. Ground water is that part of sub-surface water which comprises all the water that exists below the surface of the solid earth[7].

The water bearing strata is known as aquifer[8]. Ground water potentiality is directly related to type of aquifer, found in various geomorphic and geologic units of the area. Therefore, it is necessary to study the geomorphology and geology of the area in relation of ground water resources.

The principal source of ground water recharge to the area is through monsoon rainfall. Additional sources of recharge are through winter rainfall, return flow from irrigation water, seepage from surface water sources and influent seepage from the streams which all contribute in replenishing the ground water body. Out of the total rainfall received by the area, only 15 percent of rainfall is infiltrating to groundwater body. The average rate of rainfall infiltration ranges from 15 - 20 percent.

Groundwater occurs under water table condition in the shallow zone of quaternary sediments as well as in the zone of weathering and fracturing of hard rocks. The exploratory drilling carried out in alluvial area has revealed confined to semi-confined conditions. Exploratory drilling in the Sambhar lake bed has encountered ground water under free flowing conditions.

The major water bearing formations in the district are (i) unconsolidated formations and (ii) hard rocks. Quaternary alluvium and blown sand form the most important water bearing formations in the Jaipur district. The distribution of the aquifers, their properties and yield of these bearing water formations are described in table 4.2 and Hydrogeological Map in figure 4.3 and Hydrogeological Cross Section in figure 4.3.

4.6.1 Unconsolidated Formations

In the greater part of the Jaipur district groundwater occurs in the unconsolidated quaternary sediments consisting of fine to medium grained sand, silt, clay and kankar. The groundwater movement takes place through the pore spaces between the grains. Groundwater occurs under unconfined to confined conditions. The confined or

Table 4.2 : Hydrogeological Conditions in Jaipur District

Hydrogeological Units	Description of the unit	Occurrence	Ground Water Flow
Younger Alluvium (Quaternary)	It mainly includes wind blown sand, talus and scree deposits with some fluvial deposits along drainage channels. Alluvium is composed of fine to medium grained sand, silt, clay and kankar in varying proportions. The deposits on the flank of hills is consisted of fine to coarse grained sand and angular fragments of rocks. Thickness of alluvium varies considerably. It generally increases northward and in major part of the area noticed less than 100 m.	The litho unit, leaving aside some southern peripheral blocks like Chaksu, Phagi, Dudu and part of Sambhar block; occupies major part of the area.	There is significant variation in direction of ground water flow. In northern part, it is SW to NE. Near Shahpura, there is a ground water basin. In other part, ground water flow has been inferred N to S with variation from NNE to SSW or NNW to SSE. Hydraulic gradient in southern part and near Shahpura varies from 1.33 to 13.33.m/km.
Older Alluvium (Quaternary)	It includes fine to medium grained sand, silt, clay and kankar in varying proportions. Thickness of alluvium generally encountered less than 50 m.	The litho unit covers entire Sanganer block and spreads in major part of Sambhar, Phagi, Dudu, Chaksu and Bassi blocks.	
Quartzite (Delhi Super Group)	The litho unit is generally of grey colour but fawn, buff and white colours have also been found in the area. Quartzite is medium to coarse grained and varies from feldspathic grit to sericitic quartzite.	The litho unit occurs as small localised pocket in Amber, Viratnagar, Bassi, Jamwa Ramgarh and Kotputli blocks.	
Phyllite and Schist, Granite Gneiss (Bhilwara Super Group)	Phyllite and schist included argillaceous meta sediments. Granite and gneiss characteristically have gneissic structure comprising light coloured feldspathic and dark ferromagnesian minerals.	Phyllite and schist occupy small area in southern peripheral part in Dudu, Jamwa Ramgarh and Phagi blocks. Granite & gneiss cover extensive area in Dudu block and extends in western peripheral part of Phagi block.	

Source : State Remote Sensing Application Centre, Govt. of Rajasthan, Jodhpur.

semi-confined conditions are caused due to the presence of silt and clay beds and kankar horizon overlying aquifers at depths. The unconsolidated formations are further subdivided into two subheads as follows:

4.6.1.1 Talus and Scree Deposits

These occur on the flanks of almost all the hills and form a potential aquifer on account of lithological composition and topographic disposition. The aquifer material consists of fine to coarse grained sand and angular pieces of rocks with varying amount of losses etc. They receive the runoff from hills. The width of the aquifer zone along the foothills varies from a few metres to over 500 m.

The width is more where the hill ranges are tectonically affected and disrupted by faulting. The talus and scree deposits form potential aquifer along the foot hills in the area towards northeast of Bassi along Bassi Banskho fault zone and along the foothills of the area in Jamwa Ramgarh block. Exploratory drilling at Banskho and Patoli (fall just outside Jaipur district) have revealed the thickness of talus and scree deposits to vary from 40 in to 55 m of which the aquifer comprises 28 to 36 m.

Groundwater occurs under semi-confined to confined conditions. Yield of dugwells vary from 100 m^3/d to 600 m^3/d. As per the testing of tubewells carried out at Banskho and Patoli, specific capacity was found to be 36 and 72.66 LPM/m, transmissivity was found to be 724.0 m^2/d and 413.017 m^2/d, and permeability was found to be 20.10 and 14.10 m/d respectively.

4.6.1.2 Alluvium and Blown Sand Deposits

In the greater part, the aquifer comprises of silt, and sand, with varying amount of Kankars. The coarser material such as gravel and, pebbles forms aquifer only in limited areas, viz; Sanganer, Ambabari and Bajaj Nagar (Jaipur), Kalyanpura, Mohana, Chandlai, all in Morel basin.

The presence of clay and silt at various horizons give rise to semi-confining to confining conditions. Kankar is found to be present invariably throughout the alluvial section of the area. The presence of the secondary calcareous deposition reduces the porosity of the formation to certain extent.

Groundwater occurs under unconfined conditions in the shallow

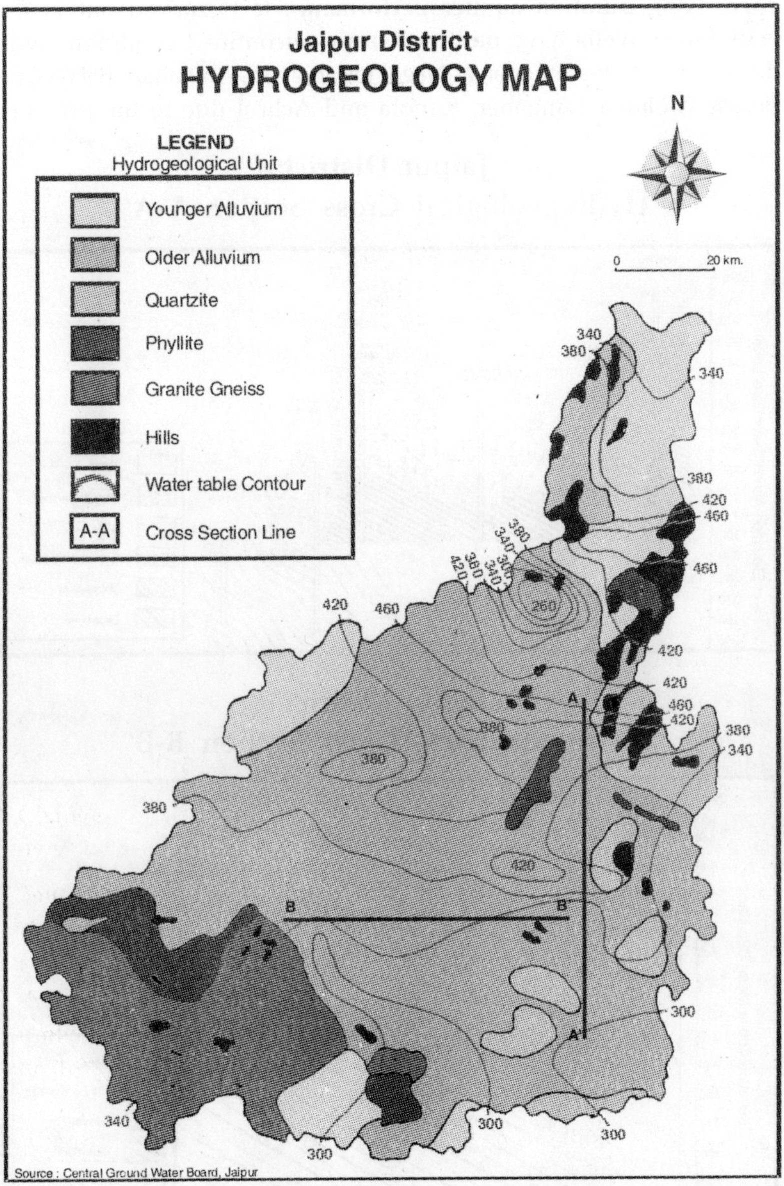
Jaipur District
HYDROGEOLOGY MAP
N
LEGEND
Hydrogeological Unit
Younger Alluvium
Older Alluvium
Quartzite
Phyllite
Granite Gneiss
Hills
Water table Contour
A-A
Cross Section Line
0
20 km.
340
380
340
380
420
460
460
380
340
300
420
420
460
260
420
420
460
420
380
340
380
380
380
420
B
B'
A
A'
300
340
300
300
300
Source : Central Ground Water Board, Jaipur

Figure : 4.3

zone. Long duration aquifer performance tests carried out in the exploratory wells have indicated, semi-unconfined conditions with delayed yield phenomenon at Nagal Jeisabohran, Ambabari, Bajajnagar, Belwa, Mohana, Sanganer, Kanota and Achrol due to the presence

Jaipur District
Hydrogeological Cross Section A-A′

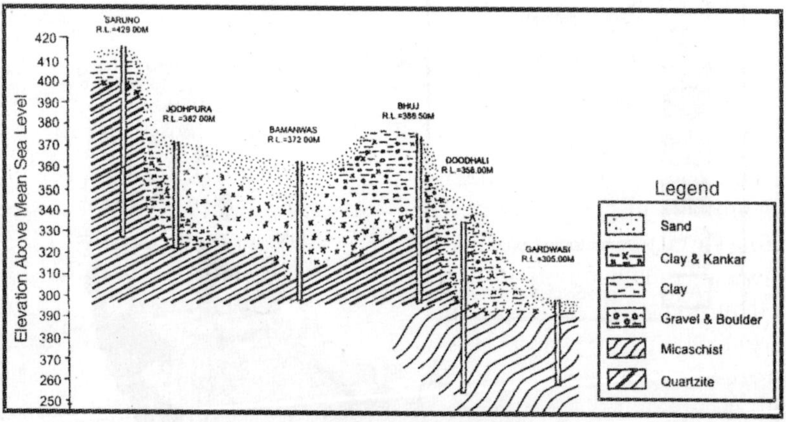

Jaipur District
Hydrogeological Cross Section B-B′

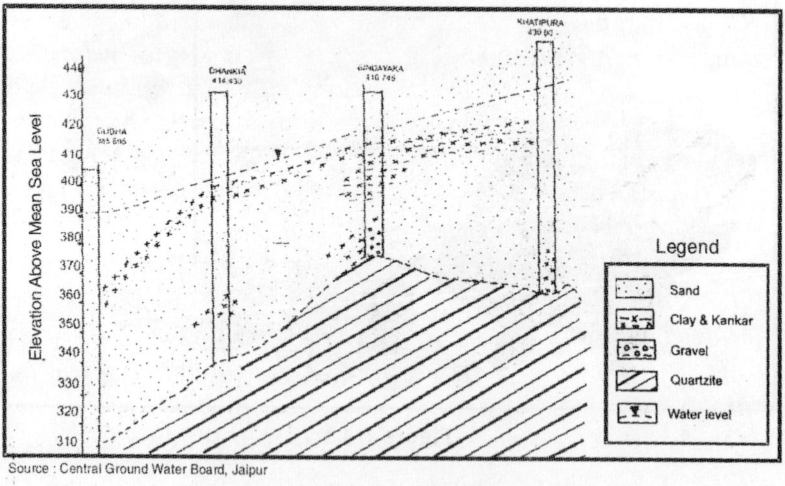

Source : Central Ground Water Board, Jaipur

Figure : 4.4

of silt and silty clay layers at different horizons. Leaky confined conditions has been observed at Tigaria in Chomu tehsil. Barrier boundary conditions are found at Jamroli and Kotkhawada in Chaksu tehsil which are due to the presence of shallow basement in the former and presence of finer sediment deposits in the vicinity of the latter. Confined conditions are reported around Kotputli and Sarund (in Kotputli tehsil) area where water gushes out with great pressure through borings in dug-cum-bored wells.[9]

The yield of wells tapping alluvial deposits vary widely and the high yield zone in the unconsolidated formations are generally related to abandoned channels of rivers and their tributaries.

In the area south of Phulera town covering Kheriram and other villages dugwells have an average depth of about 20 m tap blown sand as aquifer and can sustain pumping continuously for 24 hours attaining pumping water level at about 13.5 m to 14 m below ground level. In the area towards north of Jobner comprising Bhojpura and other villages dugwells have 19 to 20 m depth and boring to a depth of 5 to 10 m. These wells tap blown sand as aquifer and sustains pumping continuously, for 24 hours without going dry.

In the area around Sanganer, dugwells and dug-cum-bored wells can yield about 50 m^3/hr. for negligible draw down. In the area covering Motukabas, Tatiawas and other villages located on the Bandi river bank (Jaipur-Chomu road) dug-cum-bored wells, have depths of 18 m to 45 m, and tap fine sand. Kankar and little clay have not touched the basement and can sustain pumping continuously for 24 hours with small draw downs. Similar condition exists in the area around Keslahala and Sirsi (Jhotwara block), Govindgarh and Mohana (Sanganer block). A tubewell at Nowas (Jhotwara block) to a depth of 36.57 m met with fine to medium grained sand with kankar and has not touched the basement.

4.6.2 Hard Rock Aquifer

Hard rock forms aquifer in the south and south western part of the district and comprises of granitic gneiss, quartz mica schist, phyllite, quartzite, pegmatite's and intrusive granites. Hard rock forms aquifer in the area covering larger part of Dudu, Phagi and Chaksu blocks. *In hard rocks, the weathered zone below the water table acts as a good storage for groundwater.*

Table 4.3 : Geomorphology description of Jaipur District (2000-2001)

Landform Units	Lithology/Material/Description	Occurrence in district	Land use /Land cover
Fluvial Origin			
Alluvial Plain	Mainly undulating land scape formed due to fluvial activity, consists of gravels, sand, silt and clay. Terrain mainly undulating, produced by extensive deposition of alluvium.	Entire southern boundary, North of Bassi, surrounding Chomu, Shahpura & Kotputli town, west of Kanota village i.e. along river Dhundh.	Double crop, single crop (Rabi / Kharif), fallow, open scrub.
Alluvial Plain (Sandy)	Flat to gentle undulating plain formed due to fluvial activity, mainly comprised of gravels, sand, silt and clay with unconsolidated material of varying lithology, predominantly sand along river.	Mainly concentrated in central and western part of district.	Marginal double crop, single crop (Kharif) open scrub fallow.
Valley Fill	Formed by fluvial activity, usually at lower topographic locations, comprising of boulders, cobbles, pebbles gravels, sand, silt and clay. The unit has consolidated sediment deposits.	Marginally in eastern part between hills.	Single crop (Rabi)
Palaeochannel	Mainly burried on abandoned stream/river comprising of coarse textured material of variable size.	West of Chomu town & north of Sabrampura.	Sandy area.
Salt Encrustaion/ Playa	Topographical depressions comprising of clay, silt sand and soluble salts, usually undrained and devoid of vegetation.	South of Sambhar lake.	Salt quarry and salt waste.
Ravine	Small, narrow, deep, depression, smaller than gorges, larger than gulley, usually carved by running water.	Wind ward side of hills in eastern part, south east of Phulera town.	Single crop (Kharif), fallow, open scrub.
Flood Plain	The surface or strip of relatively smooth land, adjacent to a river channel, fomed by river and covered with water when river over flows its bank. Normally subject to periodic flooding.	Along rivers Dhundh & Mendha.	Single crop (Rabi), fallow, open scrub.

Contd...

...Contd.

Landform Units	Lithology/Material/Description	Occurrence in district	Land use /Land cover
Denudational Origin			
Pediment	Broad gently sloping rock flooring, erosional surface of low relief between hill and plain, comprised of varied lithology, criss crossed by fractures & faults.	Along hills in eastern and northern part of district also west of Sanganer town.	Single crop (Kharif)
Burried Pediment	Pediment covered essentially with relatively thicker alluvial, colluvial or weathered materials.	Mainly scattered in north and north east.	Single crop (Kharif)
Intermontane Valley	Depression between mountains, generally broad & linear, filled with colluvial deposits.	Marginally in between hills near Viratnagar.	Marginal double crop.
Eolian Origin			
Sandy Plain	Formed by aeolian activity, wind blown sand with gentle sloping to undulating plain, comprising of coarse sand, fine sand, silt & clay.	South of Sambhar lake, east of Hingonia Sagar.	Single crop (Kharif), land with or without scrub.
Eolian Plain	Formed by aeolian activity, with sand dunes of varying height, size, slope. Long stretches of sand sheet. Gentle sloping flat to undulating plain, comprised of fine to medium grained sand and silt. Also scattered xerophytic vegetation.	South of Kotputli town.	Single crop (Kharif) fallow, open scrub.
Hills			
Denudational Hill	Steep sided, relict hills undergone denudation, comprising of varying lithology with joints, fractures and lineaments.	East & north east of Jaipur city, around of Jamwa Ramgarh lake.	Open scrub, forest, mining.
Structural Hill	Linear to arcuate hills showing definite trend-lines with varying lithology associated with folding, faulting etc.	Scattered in northern and eastern part.	Forest, open scrub, mining.
Linear Ridge	Long narrow low-lying ridge usually barren having high run-off may form over varying lithology with controlled strike.	Scattered in eastern part.	Barren, with or without scrub, forest, ming.

Source : State Remote Sensing Application Centre, Govt. of Rajasthan, Jodhpur.

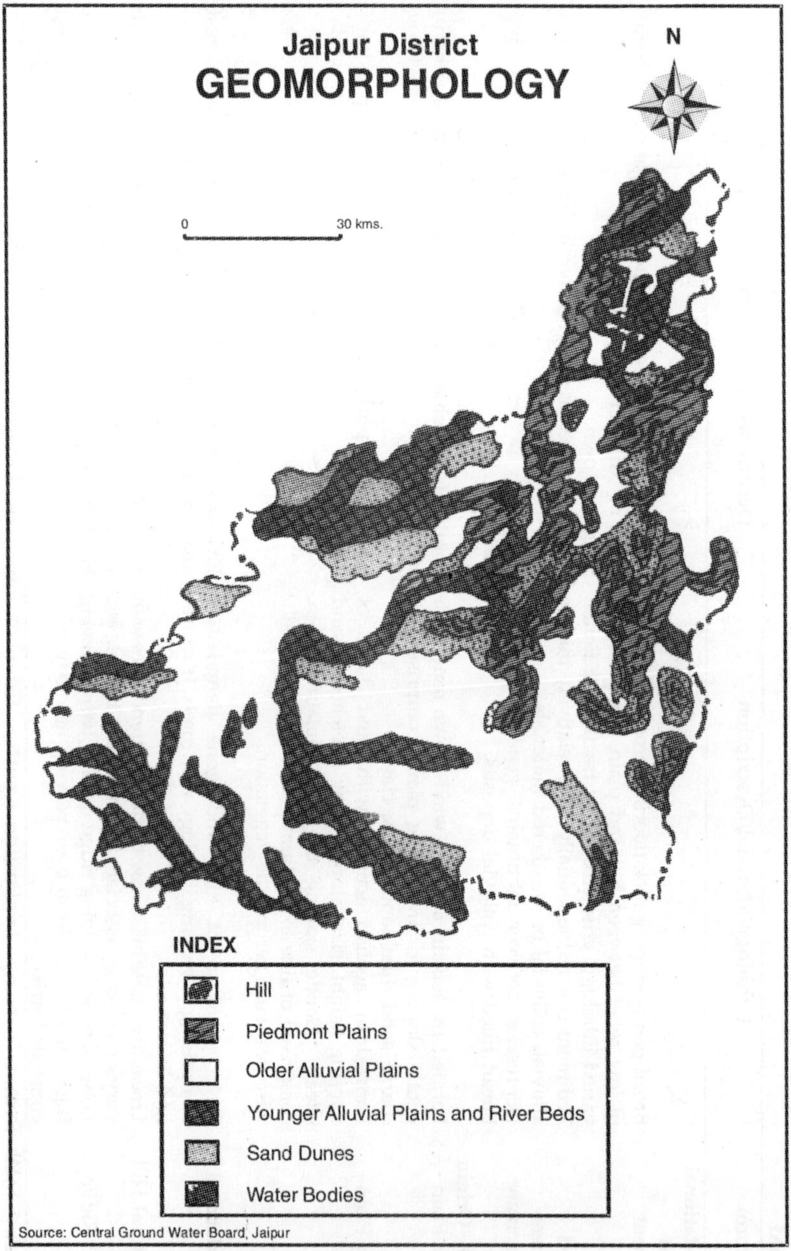
Jaipur District
GEOMORPHOLOGY
N
0
30 kms.
INDEX
Hill
Piedmont Plains
Older Alluvial Plains
Younger Alluvial Plains and River Beds
Sand Dunes
Water Bodies
Source: Central Ground Water Board, Jaipur

Figure : 4.6

The movement of groundwater is further controlled by the extent, size, openness, continuity and interconnection of fractures. Foliation joints though present, are very thin and close with death. Groundwater occurs under unconfined conditions in the shallow zone the depth of weathering has been reported to vary widely from 2 m to 30 m. The exploratory drilling in the alluvial area has revealed the thickness of underlying weathered zone to vary from 2.49 in (Ramgarh Pachwara) to 18.61 in (Kotkavda).

The specific capacity of dugwells at Dudu tapping schist and gneiss and Chittora-Renwal tapping schist has been reported to be 26 lpm/m and 16 Ipm/m respectively. The optimum yields of the same were found to be 3971 lpm and 3972 lpm respectively. Slug test conducted by C.G.W.B Jaipur, in a borewell at Chittora Renwal having a depth of 60 m with gneiss as aquifer has revealed the transmissivity and storage coefficient values to be 7.5 m^2/d and 1 x 10-2 respectively.

4.7 GEOMORPHOLOGY

From the water resources study point of view seven broad geomorphic units have been recognised in the study area using landsat imagery (Figure 4.5). Geomorphology is the most significant in ground water assessment because the rate of percolation run-off, discharge and recharge vary in each geomorphic unit; therefore ground water potentiality also varies. Geomorphology description of Jaipur district show in table 4.3.

In hilly region exploration of under ground water is limited because of high construction cost of wells, though very high water accumulation is found there due to multi phased tectonic activity. In piedmont plains medium to low water potentiality is found due to the presence of hard rock beneath. Older alluvial plains are found throughout the district with medium water potentiality. In younger alluvial plains ground water potentiality is very high because coarse to medium textured sands to loamy sand and sandy loam soils are good water bearing strata. Besides this, plains are found along the river course and rivers flow along water bearing lineaments or faults (Table 4.3). Generally sand dunes have very low water potentiality but some old sand dunes have low water potentiality.

4.8 DEPTH OF WATER TABLE

Water levels are being monitored by "Centre Ground Water

Board, Jaipur" in 68 national hydrograph network stations selected in different types of hydrogeological formations in the district.

Depth of water level ranges from less than 5 metre below the ground level (bgl) to 45 mbgl. In major part of the Jaipur district depth to water level is within 20 mbgl. Shallowest water level i.e. less than 5 m has been observed in the eastern part of the district around Jatwara railway station, around Jalmahal, alluvial tracts of Bandi river, some places in alluvial tracts of Sabi and Banganga river basins and also in the south western part of the district, around Mangarwara and Mozamabad.

Deepest water level has been monitored at Chomu (More than 45 mbgl). In Jaipur tehsil there is wide range in depth of water table and its ranges from less than 5 mbgl (around city) to more than 33 mbgl (around Jhotwara).

4.9 SLOPE OF THE WATER TABLE

The elevation of water table ranges from around 296m amsl (in the southern part of district) to around 478m amsl (in the north western part of the district). In Jaipur tehsil elevation of water table is around 400 m amsl. In the western part of the district water level slopes towards south western direction and in the south western part of the district before crossing Bandi river water table slopes towards south western direction and after the crossing, the river slope direction becomes south easterly. In southern part of the district, slope direction of the water table is towards south.

4.10 WATER POTENTIAL ZONE

To delineate water potential zones in the district, the maps of Geology (In chapter-2 Figure 2.3), Geomorphology (Figure 4.6), table 4.3 and table 4.4 were superimposed and considering all these factors simultaneously, water potential zones have been demarcated (Figure 4.7). Similarly technique was adopted by Ramaswamy *et.al.*[10], Chaturvedi et.al.[11], Dewangan et.al.[12] and Balasubramaniam *et.al.*[13].

4.10.1 Very High Water Potential Zone

Table 4.4 reveals that about 4383.51 Km^2 area of district has very high water potentiality. Particularly the north and western part of the district falls under this zone. Small patches of this zone are found in Kotputli, Viratnagar, Sambhar, Govindgarh and Sanganer tehsils. Geomorphically this zone spreads over younger alluvial plains

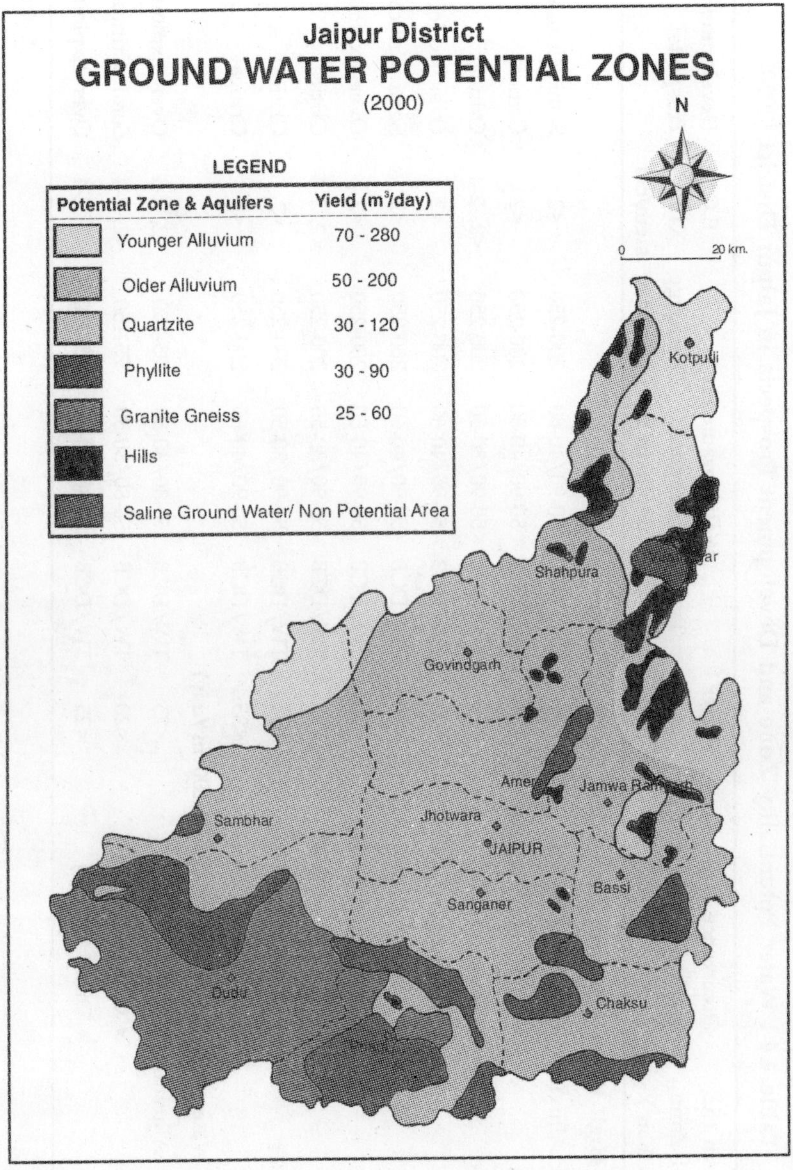
Jaipur District
GROUND WATER POTENTIAL ZONES
(2000)
N
0
20 km.
LEGEND
Potential Zone & Aquifers
Yield (m³/day)
Younger Alluvium
70 - 280
Older Alluvium
50 - 200
Quartzite
30 - 120
Phyllite
30 - 90
Granite Gneiss
25 - 60
Hills
Saline Ground Water/ Non Potential Area
Kotputli
Shahpura
Govindgarh
Amer
Sambhar
Jhotwara
JAIPUR
Bassi
Sanganer
Dudu
Chaksu

Figure : 4.7

Table 4.4 : Water Potentiality Zone and Development Prospects in Jaipur District (2000)

Aquifer in the Potential Zone (Area in Km^2)	Occurrence Block (Area in Km^2)	Water Level (m.)	Well Parameters Type	Proposed depth in m.	Discharge in m^3/day	E.C. X10^{3}µ siem/cm	Development Prospects
1. Very High Water Potential Zone (Yield 70-280 m^3/day)							
Younger Alluvium (4383.51)	• Amber (746.46)	<35	TW/DCB	50-90/40-60	200-250	<2	Semi Critical
	• Viratnagar (588.38)	<30	TW/DCB	50-90/40-60	200-250	<2	Critical
	• Bassi (218.85)	<15	TW/DCB	50-90/30-50	200-250	<2, 2-4	Critical
	• Govindgarh (638.08)	<30	TW/DCB	50-90/40-60	200-250	<2	Over Exploited
	• J. Ramgarh (577.82)	<35	TW/DCB	50-90/40-60	200-250	<2	Semi Critical
	• Jhotwara (553.04)	<35	TW/DCB	50-90/40-60	200-250	<2	Over Exploited
	• Kotputli (427.32)	<20	TW/DCB	50-90/30-50	200-250	<2, 2-4	Critical
	• Sambhar (249.44)	<15	TW/DCB	50-90/30-50	200-250	<2	Critical
	• Shahpura (384.14)	<35	TW/DCB	50-90/40-60	200-250	<2	Critical
2. High Water Potential Zone (Yield 50-200 m^3/day)							
Younger Alluvium (2828.96)	• Bassi (357.62)	<25	TW/DCB	50-80/30-50	75-150	<2, 2-4	Over Exploited
	• Chaksu (552.06)	<20	TW/DCB	50-80/30-50	75-150	<2, 2-4	Semi Critical
	• Dudu (327.26)	<25	TW/DCB	50-80/30-50	75-150	<2, 2-4	Over Exploited

Contd...

...Contd.

Aquifer in the Potential Zone (Area in Km^2)	Occurrence Block (Area in Km^2)	Water Level (m.)	Well Parameters Type	Proposed depth in m.	Discharge in m^3/day	E.C. X$10^3\mu$ siem/cm	Development Prospects
	• Phagi (396.11)	<30	TW/DCB	50-80/40-60	75-150	<2, 2-4	Over Exploited
	• Sambhar (582.04)	<25	TW/DCB	50-80/30-50	75-150	<2	Over Exploited
	• Sanganer (613.87)	<35	TW/DCB	50-80/40-60	75-150	<2	Over Exploited
3. Medium Water Potential Zone (Yield 30-120 m3/day)							
Quartzite (540.72)	• Amber (104.02)	<35	Tw/DW	45-70/40-50	40-100	<2	Over Exploited
	• Viratnagar (75.60)	<10	Tw/DW	45-70/30-50	40-100	<2	Over Exploited
	• Bassi (54.76)	<25	Tw/DW	45-70/35-50	40-100	<2, 2-4	Over Exploited
	• J.Ramgarh (175.80)	<20	Tw/DW	45-70/30-50	40-100	<2	Over Exploited
	• Kotputli (130.54)	<25	Tw/DW	45-70/35-50	40-100	<2, 2-4	Over Exploited
4. Low Water Potential Zone (Yield 30-90 m3/day)							
Phyllite and	• Chaksu (177.30)	<15	Tw/DW	50-60/20-25	40-90	<2, 2-4	Semi Critical
Schist (704.64)	• Dudu (407.93)	<30	Tw/DW	50-60/35-50	40-90	<2, 2-4	Safe
	• J. Ramgarh (136.01)	<20	Tw/DW	50-60/30-50	40-90	<2	Semi Critical
	• Phagi (73.40)	<15	Tw/DW	50-60/30-50	40-90	<2, 2-4	Semi Critical

Contd...

...Contd.

Aquifer in the Potential Zone (Area in Km^2)	Occurrence Block (Area in Km^2)	Water Level (m.)	Well Parameters Type	Proposed depth in m.	Discharge in m^3/day	E.C. $X10^3\mu$ siem/cm	Development Prospects
5. Very Low Water Potential Zone (Yield 25-90 m3/day)							
Granite Gneiss (1431.93)	• Dudu (1055.68)	<20	Tw/DW	50-70/30-50	40-80	<2, 2-4	Critical
	• J. Ramgarh (72.25)	<10	Tw/DW	50-70/25-40	40-80	<2	Safe
	• Phagi (304.00)	<20	Tw/DW	50-70/30-50	40-80	<2, 2-4	Safe

Index : **TW** - Tube wells, **DCB** - Dug cum borewells, **DW** - Dug wells **Safe** - <65% stage of development
Semi Critical - 65-85% stage of development, **Critical** - 85-100% stage of development,
Over Exploited - >100% stage of development,

Source : State Remote Sensing Application Centre, Govt. of Rajasthan, Jodhpur.

between the hilly ranges. Agricultural practice is intensive only on younger alluvial plains. In this zone average yield of well is 200-250 m^3/day.

4.10.2 High Water Potential Zone

This zone constitutes about 2828.96 Km^2 area of the district. This zone passes through southern-east, north-west and central part of the district. In this zone younger alluvial plains of river Bandi, Mashi and Morel are found along with the high lineament intersection density and intensive agricultural practices. The average yield of well of this zone is 75-150 m^3/day (Table 4.4).

4.10.3 Medium Water Potential Zone

About one third area of the district belongs to medium water potential zone around 540.72 Km^2. This zone passes through north-east border of the district. Some small patches are found in Amber, Viratnagar, Jamwa Ramgarh, Bassi and Kotputli tehsils. The average yield of wells in this zone is 40-100 m^3/day (Table 4.4).

4.10.4 Low Water Potential Zone

Fourth zone constitutes about 704.64 Km^2 area of the district and has low water potentiality. Mainly four patches are found in Chaksu, Dudu, Phagi and J. Ramgarh tehsil. In these area generally rain-fed crops are sown on calcareous soils. The average yield of well of this zone is 40-90 m^3/day (Table 4.4).

4.10.5 Very Low Water Potential Zone

Fifth zone constitutes about 1431.93 Km^2 area of the district and has very low water potentiality. This patch is surrounded by sand dunes in the western parts of the district, mainly in Dudu, Phagi and Jamwa Ramgarh tehsil. In these zone generally rain-fed is sown. The average yield of well of this zone is 40-80 m^3/day (Table 4.4).

REFERENCES

1. Kuiper, E. (1971) : *Water Resource Development*, Butterwoths, London, p. 5.
2. Saini, R.R. & Kalwar, S.C. : *Remote Sensing in Resource Geography*. Pointer Publication, Jaipur, p. 54.
3. Van Te Chow (1964) : *Hand Book of Applied Hydrology*. Mc Graw Hill Book Company, New York, pp. 1-2.

4. Gardner, J.S. (1971) : *Physical Geography*. Harper College Press, London, p. 154.

5. Van Te Chow (1964) : *Hand Book of Applied Hydrology*. Mc Graw Hill Book Company, New York, p. 617.

6. Van Te Chow (1964) : *Ibid*, p. 368

7. Hussain, S. R. (1965) : *Text Book of Water Supply and Sanitation Engineering*, Oxford & USA Pub. Co., New Delhi, p. 93.

8. Ward, R. C. (1978) : *Ground Water*, Chapman and Hall Company, New York, p. 19.

9. Siddiqui (1970) : *General Report-Geological Survey of India*, p. 76.

10. Ramaswamy, S. M. & Bakliwal, P. C. (1983), "Ground Water Targeting of Banded Gneisses Complex, Rajasthan Through Remote Sensing, Proc. Nat. Symp. on Remote Sensing in Development and Management of Water Resources.

11. Chaturvedi, R. S. et. al (1985) : "Integrating Remote Sensing Technology in Ground Water Exploration-" Typical Case Study from Bundelkhand Region in U.P. Proc. Nat. Symp. on Remote Sensing in Development and Management of Water Resources.

12. Dewangen, K. N. & Shanker, U. S. (1983) : "Use of Landsat Imagery in Ground Water Investigation in Part of Noyil Basin, Tamil Nadu," Proc. Nat. Symp. on Remote Sensing in Development and Management of Water Resources.

13. Balasubramaniam, T. (1983): "Application of Remote Sensing to Ground Water Studies-A Case Study from North Arcot District of Tamil Nadu, Proc. Nat. Symp. on Remote Sensing in Development and Management of Water Resources.

5

Major Problem of Water Quality

The quality of water is of almost equal importance to quantity in any water supply planning, and especially for drinking purposes purity is absolutely essential, and for stock purposes also a pure supply is almost equally important. Although the quality of water varies according to purposes, the water having certain composition may be useful for a particular purpose and may not be suitable for other purpose. The chemical, physical and bacterial characteristics of water determine its usefulness for any purpose, e.g., domestic, irrigation or industrial, etc. Some important uses and major quality parameters involved in deciding its suitability for different uses are considered in Table 5.1. Specially in case of drinking water supply for human consumption, it must be of very good quality.[1]

Table 5.1 : Water uses and quality considerations[2]

Water-Use	Important Quality Parameter
Power	Dissolved oxygen, pH
Irrigation	Dissolved solids, electrical conductivity.
Potable water supply	Colour turbidity, hardness, pathogenic organisms, organics.
Fishing	Dissolved oxygen, Pesticides, $C0_2$, pH, heavy metals
Recreation	Pathogenic organism, pH

Water, whether derived from underground channels, by dug-wells, natural springs or obtained from surface water in river and reservoirs, is never chemically pure but always contains various solids and gaseous compounds in solution [3]. In fact, the water from two neighbouring sources, whether springs, river or wells, very rarely have precisely the same composition. Two waters may have the same appearance and yet be very unlike in taste and chemical character.

Safe drinking water is the primary need of every human being. Rural population mainly depends upon ground water sources. Ground water sources are not safe when they contain excess of salinity, iron, manganese, nitrate, fluoride, etc. Fluoride although beneficial when present in concentration of 0.8 to 1.0 mg/l, has been associated with mottled enamel of the teeth when present in water in excess of 1.5 mg/l. Skeletal fluorosis has been observed in humans where water contains more than 3 mg/l of fluoride.

5.1 SOURCES OF IMPURITIES

The water which percolates into the soil and appears elsewhere in the form of springs or percolation wells, absorbs other acids from decaying vegetables matter therein, (e.g., root etc.). These acids dissolve some minerals not soluble in pure waters while other minerals and salts are dissolved by the water as such.

Composition of inland and ground waters is influenced by the chemical nature of the earth's crust and the distribution of the rock types. Rocks are chemically weathered by the processes of hydrolysis, oxidation-reduction, Hydration and carbonation into soil comprising of mineral particles like quartz-zircon, tourmaline, garnets, clay and hydrated oxides of iron and aluminium besides other chemicals of soluble nature. Leaching through the soil removes soluble matter into the infiltrating water. Natural waters are comprised mainly of contains like calcium, magnesium, sodium and potassium together with the anions like bicarbonate, carbonate, sulphate, chloride and nitrate. The oxides of aluminium and iron are often present in colloidal form in water. Besides geology, the chemical nature of the waters is also influenced by the ionic composition of rain and human activities like disposal of industrial and domestic waste waters, and agricultural run-off. The growth of algae and micro-organisms also influence the quality of water to a great extent.

Bacteria and fungi, generally induce loose motion. Probably no surface water is free from them, and there may be millions in a cubic centimetre.[4] Most of them are harmless, and many assist in the decomposition of organic matter and are essential to the self-purification of water.

Another important class of living suspended microscopic organism is algae, which is found even in rain water. Although it is not injurious to health, if taken in large quantities it may temporarily causes intestinal trouble. Algae can produce odours and tastes that are decidedly unpleasant.

5.2 WATER STANDARDS

In view of the wide variations in the chemical composition of waters in different parts of the world, rigid standards of chemical quality are difficult to be established. The limits are; therefore, designated as "acceptable" or "desirable" etc. indicating the suitability of water to the best possible extent to the consumers. The values greater than those listed as "maximum allowable", "maximum permissible" or "cause of rejection" etc. will greatly impair the portability and in that case the water should normally be avoided for use.

In India, agencies like Indian Council of Medical Research (ICMR), Bureau of Indian Standards (BIS) and Ministry of Works and Housing have recommended drinking water standards, which are being followed at different places with or without modifications depending upon the limitations posed by the general water quality met within that area.[5] The standards for potable water laid down by World Health Organization (WHO) are called International standards. However, most of these standards are only indicative and can be modified or disregarded in specific instances (*Appendix* - 4).

5.3 WATER QUALITY INVESTIGATION

There are a large number of hydrograph stations which have been established all over the district by *Central Ground Water Board, Jaipur* for long term monitoring of water levels and these stations include both open wells and piezometers. From the open wells once in a year i.e. during premonsoon period water samples are collected for detailed chemical analysis. Chemical quality of ground water of the district is studied every year from data obtained through chemical analysis of these water samples.

Entire Jaipur district has been covered long back by systematic hydrogeological surveys and at present part of the district has been covered by reappraisal hydrogeological surveys by the State Ground Water Board and State Remote Sensing Department, Jodhpur. During these surveys water samples of ground water were collected and analysed for study of the chemical quality of ground water of the district.

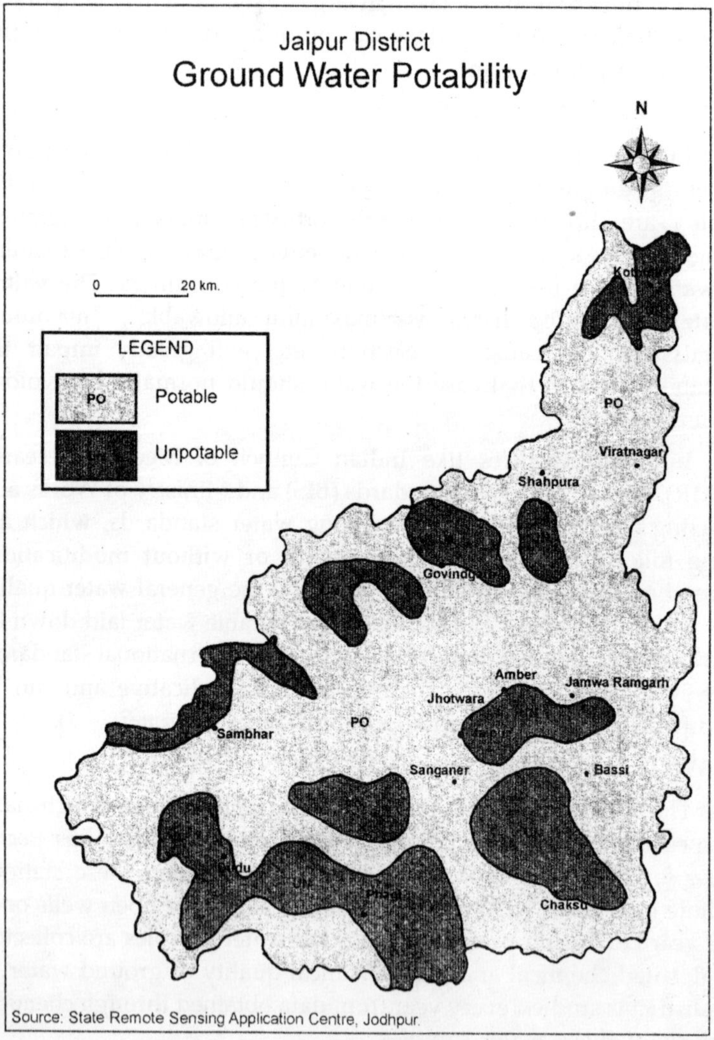

Source: State Remote Sensing Application Centre, Jodhpur.

Figure : 5.1

Table 5.2 : Result of Chemical Analysis of Water Samples Collected from various Location of Jaipur District (May 2000)

No.	Location	pH	Electrical Conductivity μS/cm at 25⁰c	CO_3	HCO_3	Cl	SO_4	NO_3	PO_4	Total Hardness as $CaCO_3$	Ca^{++}	Mg^{++}	Na^+	K^+	F	B	Fe
				Value express in ppm											Value in mg/L		
1	2	3	4	5	6	7	8	9	10	11	12	13	14	15	16	17	18
1	Hastera	7.35	2080	Nil	695	319	-	49	0.16	630	116	83	184	19	0.70	0.25	0.50
2	Amber	7.50	870	Nil	317	92	30	4	0.20	300	76	27	48	19	0.25	-	0.16
3	Bhojpura	8.10	1363	Nil	512	99	100	23	0.10	385	40	69	136	11	1.30	0.60	0.60
4	Chaksu	7.91	1480	Nil	605	92	82	20	0.40	150	24	22	299	5	2.00	0.36	0.50
5	Dudu	8.70	3370	48	415	682	308	100	0.10	380	44	66	644	7	1.45	0.80	-
6	Jaipur (GIS)	8.40	1480	12	451	170	80	220	0.10	550	92	78	156	9	0.18	-	-
7	Jhir	8.69	900	48	390	43	34	120	0.10	250	40	36	110	3	0.30	0.25	-
8	Kotputli	9.20	1210	24	500	106	75	18	0.10	200	20	36	90	3	1.00	0.32	0.59
9	Nasnota	7.78	5140	Nil	647	1086	403	200	0.18	300	40	49	1035	84	11.00	1.91	-
10	Phagi	8.60	2880	48	390	391	451	50	0.18	210	36	29	575	7	2.70	0.80	0.22
11	Phulera	8.30	2220	-	464	490	126	100	0.10	460	56	78	379	4	0.70	0.52	0.22
12	Sanganer	7.50	1320	Nil	732	35	115	20	0.55	160	12	32	298	2	7.30	0.63	-
14	Nangal turkivan (Govindgarh II)	7.75	1068	Nil	451	92	20	13	0.70	440	68	66	73	12	0.40	0.16	-
15	Jatwara Rly. Station	7.75	11110	Nil	543	1372	2800	4	0.10	1940	180	362	1841	8	2.30	1.96	-

Contd...

...Contd.

1	2	3	4	5	6	7	8	9	10	11	12	13	14	15	16	17	18
16	Ramniwas Bag (Jaipur)	8.60	2140	24	207	362	152	350	0.10	730	168	75	105	9	-	0.13	-
17	Jaipur Air Force Station	8.60	1820	12	244	426	53	40	0.10	550	96	75	166	2	1.00	0.24	0.29
18	Sisodia Rani Bag	7.90	790	Nil	323	57	30	40	3.40	295	80	23	46	8	0.47	0.15	-
19	Senicharki Bagichi (Jaipur)	7.80	3130	Nil	1110	355	246	150	3.20	900	128	141	207	312	0.40	0.16	0.45
20	Baikabas	8.91	1400	96	586	21	55	100	0.20	50	4	10	350	2	15.60	0.19	-
21	Jobner	8.40	6085	72	1025	1092	550	90	0.12	490	24	105	1219	13	1.80	0.90	-
22	Shivdaspura	8.13	1440	Nil	525	107	144	100	0.18	450	72	66	166	5	1.00	0.27	0.38
23	Kalyanji Ka Rasta (Jaipur)	8.40	1850	12	744	213	71	100	2.60	480	104	53	184	164	0.30	0.19	-
24	Jalebi Chowk (Jaipur)	8.40	1650	24	305	192	45	240	0.10	320	72	34	175	140	0.40	0.13	-
25	Mozamabad	7.06	24340	Nil	1452	7100	1934	200	0.10	4200	288	846	3818	686	0.10	4.05	0.22
26	Rupari kalan	8.61	1360	36	439	156	69	120	2.50	170	36	19	290	10	0.85	0.47	0.10
27	Dhola	7.80	490	Nil	305	21	18	5	0.10	260	72	19	21	7	0.40	0.10	0.37
28	Tunga	8.30	1440	-	232	213	140	206	0.10	350	100	24	207	5	0.50	0.44	-
29	Dawach	8.81	1480	60	622	107	148	-	2.10	330	40	56	276	3	1.80	0.19	-
30	Mangalwara	7.50	24700	Nil	512	7242	1820	-	3.10	2482	468	318	4600	34	1.80	3.75	0.22
31	Padsoli	8.17	2390	60	342	568	1160	-	0.85	204	76	22	316	5	4.30	0.81	0.31
32	Rahlana	8.45	1400	36	349	492	-	-	1.00	540	59	75	138	20	0.40	0.27	-
33	Hastal Ka Bas	7.85	7000	Nil	878	1626	461	500	3.10	980	128	165	1334	17	3.00	1.39	-
34	Pallukhurd	8.60	7490	48	927	1775	578	100	1.40	1340	80	294	1357	29	6.60	1.63	-

Source : State Ground Water Board, Jodhpur.

5.3.1 Water Potability

Based on the chemical analysis data of the water samples collected from hydrograph stations during May 2000 drinking water quality map (Potable/Unpotable) of the district has been prepared (Figure 5.1) and the data are given in Table 5.2.

On viewing the drinking water quality map based on permissible limit of nitrate, fluoride and salinity, the area in the northern part of the Jaipur district comprising Kotputli block, lower portion of Shahpura, northwest area between Sambhar and Govindgarh and southwest part of the district covering Dudu, Sambhar and Sanganer block and central part of Jamwa Ramgarh block is not suitable for drinking (Figure 5.1).

5.3.2 Salinity

The ground water in the district is mostly characterised by bicarbonate type of water (56 percent) out of which 42 percent is classified as sodium bicarbonate, 8 percent is of calcium plus magnesium bicarbonate type and 6 percent is mixed bicarbonate type. This type of water is generally fresh in nature with electrical conductivity (EC) less than 1500 μS/cm except at some places where EC of water is between 1500-3000 μS/cm. The mixed type of water constitutes only 22 percent in which 14 percent is of Na-mix, 5 percent is Ca-Mg- mix and rest 3 percent is mix-mix type of water. The EC of these water ranges between 1500-3000 mS/cm but at some places it may exceed upto 4000 μS/cm. The rest 22 percent ground water is chloride type having sodium as the dominant ion. These waters have EC ranging between 3000-12000 μS/cm.

The ground water is characterised by low to high salinity as viewed from the salinity map in figure 5.2. The fresh to slightly saline water having EC below 2000 mS/cm at 25°C occurs in most part of the district. In eastern part of the district around Hastal Ka Bas (Block-Jamwa Ramgarh), and Bassi, Jatwara Railway Station (Block-Bassi) electrical conductivity value of ground water is more than 5000 μS/cm at 25°C. In south western part of the district around Mangalwara, Mozamabad, Nasnota (Block-Dudu), Pallukhurd and Jobner (Block-Sambhar) electrical conductivity value of ground water is more than 5000 μS/cm at 25°C (Table 5.2). Two small patches of this range are one around Kotputli and the other between Shahpura and Amber block. In southwest a patch between Phagi, Dudu, Sambhar and Sanganer block also have such type of water.

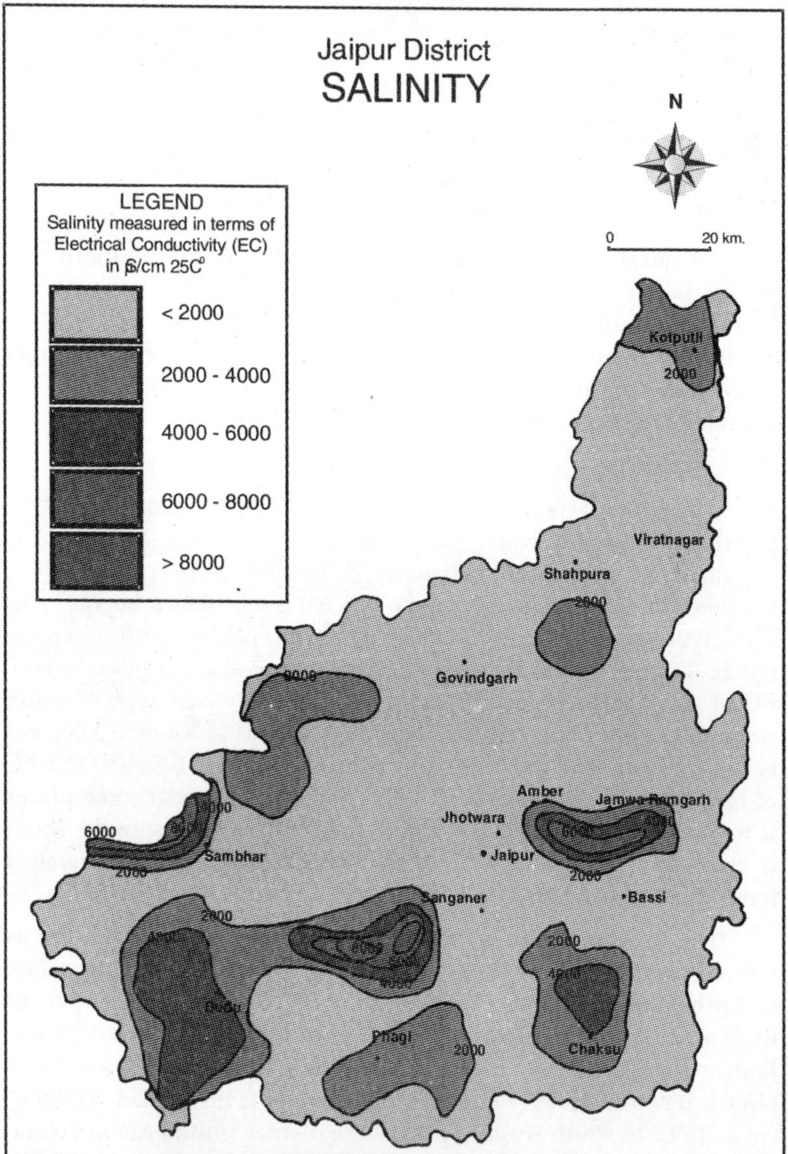

Source: State Remote Sensing Application Centre, Jodhpur.

Figure : 5.2

The ground water salinity in 0 to 2000 µS/cm range is available in 80 percent in Amber, 84 percent in Viratnagar, 100 percent in Bassi,

73 percent in Chaksu, 50 percent in Dudu, 92 percent in Govindgarh, 60 percent in Jamwa Ramgarh, 70 percent in Jhotwara, 48 percent in Kotputli, 71 percent in Phagi, 84 percent in Sambhar, 50 percent in Sanganer and 92 percent in Shahpura block and is characterised as fresh to slightly saline water. The next range of moderately saline water i.e. EC 2000 to 4000 mS/cm is represented by 20 percent in Amber, 6 percent in Viratnagar, 13 percent in Chaksu, 15 percent in Dudu, 8 percent in Govindgarh, 10 percent in Jamwa Ramgarh, 52 percent in Kotputli, 29 percent in Phagi, 31 percent in Sambhar, 25 percent in Sanganer and 8 percent in Shahpura blocks. EC 4000 and above is represented by 13 percent in Chaksu, 35 percent in Dudu, 30 percent each in Jamwa Ramgarh and Jhotwara, 15 percent in Sambhar and 25 percent in Sanganer block.

5.3.3 Chloride

Chlorides occur naturally in all types of water. The chloride concentration is within 500 Parts Per Million (ppm) in the greater part of the district, which contains excessive amount beyond the permissive limit of drinking water standard [6]. Chloride Concentration of 250 ppm or less is found in the area covered by blown sand and along the river. During systematic and reappraisal hydrogeological surveys carried out by C.G.W.B. department the chloride concentration of more than 1000 ppm was reported in the ground water at the following places.

Sambhar block

- Sambhar salt lake and its peripheral area.
- Around Bhaislana and in the vicinity of Mendha river towards south of Kishangarh.

Sanganer block

- In the area around Chirota and Sirah.

Dudu block

- In the area towards northeast, north and west of Mozamabad and between Chhaparwa bund and Narikhera. The Chloride concentration in a well north of Mozamabad is reported to be 6940 ppm.
- In a patch comprising the area within Dudu in the east, Naraina in north, Sakhun in west and Chhaparwa bund in south.

Jamwa Ramgarh block

- West of Arwari Tal and around Dugrawata

Sanganer and Bassi block

- In the area of Amanisha nala, Dhund and Ratanganga rivers.

Bassi block

- Around Nangl Bohra, Laxmi puri, kherla and Kameti.

Chaksu block

- Towards north of Chaksu town, also south of Chaksu railway station and area around Ralaotha.

Kotputli block

- Around Chardera and Dhani Jhera.

5.3.4 Nitrate

Nitrate in water is considered to be a part of the final oxidation product of nitrogenous organic matter. Studies have shown that nitrate (NO_3^-) in excess of about 44 ppm in drinking water may be a contributing factor or the cause of "methemoglobinema" (blue bodies) among infants.[7]

The map of nitrate distribution in figure 5.3, the Jaipur district shows that higher concentration of nitrate i.e. more than 100 mg/L is mostly observed in northern part of the district including Kotputli block and two patches in Govindgarh block. Other places where such quality of water is encountered in the district are around villages Manda, Hingonia, Gadota, Untirampura and Bagru near Jaipur. The area in southwest around Dudu, in eastern side around village Nithara, Saiwad and Shahpura in Shahpura block, Bhanpurakallan, Andhi, Bhoojdhanlu, Rasala in Jamwa Ramgarh and in south-eastern part in Bassi and Chaksu block is having nitrate in the range 51 to 100 mg/L. Water with nitrate concentration 0 to 50 mg/L is found in rest of the district.

The distribution pattern of nitrate concentration shows that 37 percent samples fall in 0 to 50 mg/L range, 42 percent in 51 to 100 mg/L range and rest 21 percent sample fall in more than 100 mg/L of nitrate. The maximum nitrate value is observed at village Untirampur (2640 mg/L) in Sanganer block.

5.3.5 Total Hardness

The district is mostly characterised by ground water having hardness ranging between 0 to 300 mg/L and constitutes 81 percent

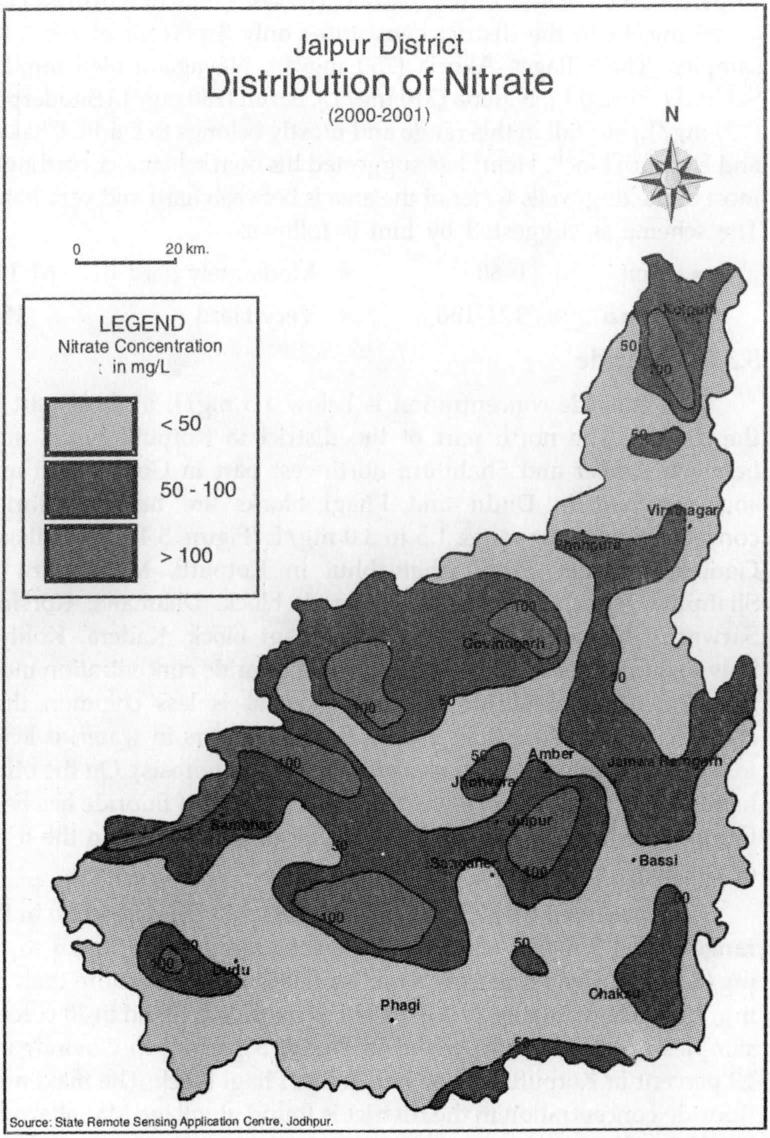

Source: State Remote Sensing Application Centre, Jodhpur.

Figure : 5.3

of total samples. Blocks Viratnagar, Bassi, Govindgarh, Kotputli, Phagi, and Shahpura are having maximum number of samples in this range. The next range of hardness of 301 to 600 mg/L is found in 16 percent well waters of the district. The last range of hardness (TH > 600 mg/L) in the district constitutes only 3 percent of the total samples. The villages Akoria (750 mg/L), Nayagaon (845 mg/L), Sakhun (780mg/L), Narena (710 mg/L), Sarun (780 mg/L) Sunderpur (705 mg/L) etc. fall in this range and mostly belongs to Dudu, Chaksu and Kotputli block. Hem[8] has suggested his own scheme, accordingly most of the dug-wells water of the area is between hard and very hard. The scheme as suggested by him is follows:

- Soft 0-60
- Moderately hard 61-120
- Hard 121-180
- Very Hard 180.

5.3.6 Fluoride

The fluoride concentration is below 1.5 mg/L in most part of the district. The north part of the district in Kotputli block, area between Amber and Shahpura northwest part in Govindgarh and southern part in Dudu and Phagi blocks are having fluoride concentration in the range 1.5 to 3.0 mg/L (Figure 5.4). The villages Gonera, Ponyala, Putli, Chaturbhuj in Kotputli, Mohanpura in Shahpura block, Tigria in Govindgarh block, Dhamana, Korsina, Sarwad, Chatyali, Phagi, Choru in Phagi block, Kadera, Kohlye, Shivdaspura in Chaksu block are having fluoride concentration more than 3.0 mg/L. Fluoride in ground water is less common than chloride. When more than 2 ppm fluoride occurs in water, it leads to the dental defect known as *mottled enamel* (Fluorosis). On the other hand, water containing between 0.7 and 1.5 ppm of fluoride has been found to reduce greatly the incidence of dental cavities in the teeth of children.[9]

It is observed that 75 percent of samples in the district lie in the range 0 to 1.5 mg/L whereas 17 percent samples are in 1.5 to 3.0 mg/L range. Only 8 percent samples fall in the range more than 3.0 mg/L. The higher range (> 3.0 mg/L) of fluoride is found in 20 percent samples in Chaksu, 15 percent in Dudu, 8 percent in Govindgarh, 19 percent in Kotputli and 35 percent in Phagi block. The maximum fluoride concentration in the district is found at village Manoharpura in Shahpura block.

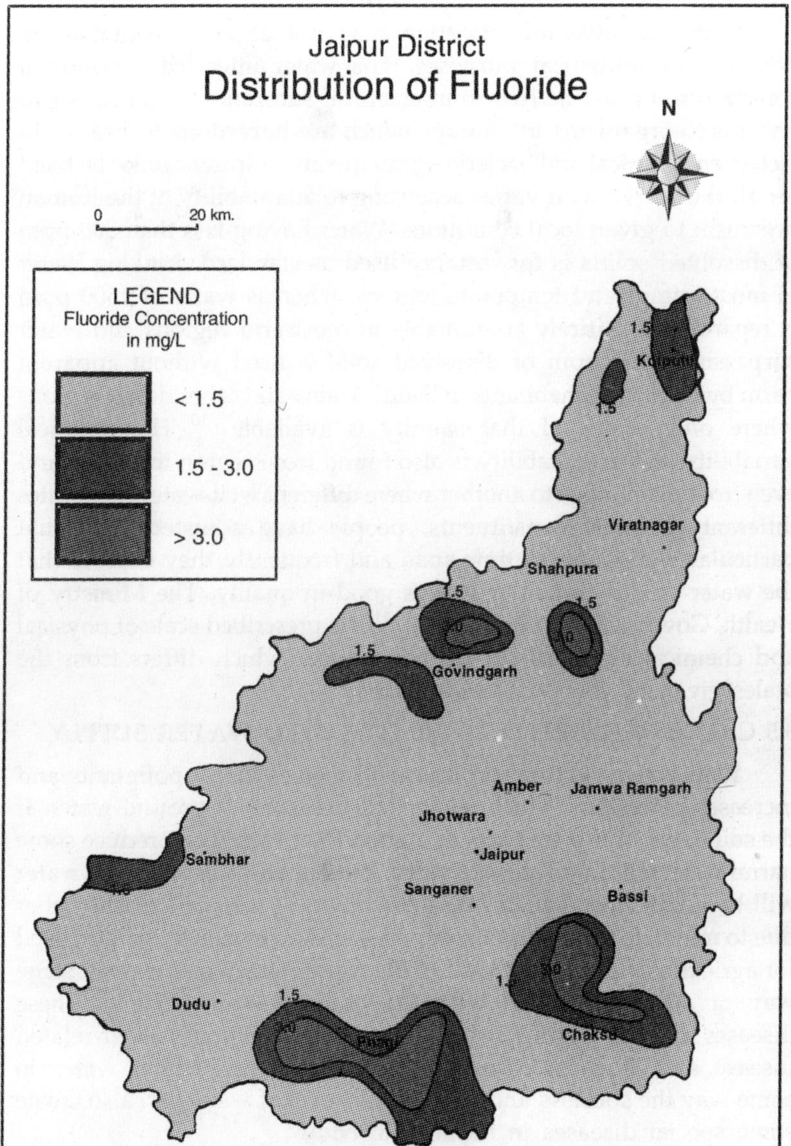

Source: State Remote Sensing Application Centre, Jodhpur.

Figure : 5.4

5.4 SUITABILITY OF WATER FOR DRINKING PURPOSES

Water suitable for irrigation may not at all be suitable for domestic or industrial purposes. The water intended for human consumption must be free from chemical substances, Guinea worm and micro-organisms in amount which are hazardous to health. In fact, even chemical and bacteriological quality of water cannot be fixed for all the world, as it varies according to adaptability of the human organism to given local conditions. Water having less than 500 ppm of dissolved solids is for instance used as standard drinking water in most humid and temperate regions, whereas water of 1500 ppm is regarded as entirely accountable in most arid regions and water surpassing 3000 ppm of dissolved solid is used without apparent harm by the local inhabitants in Saudi Arabia and other desert regions where only water of that salinity is available. [10] The chemical variability and acceptability is also found from region to region and even from one village to another where different well-water constitutes different chemical constituents; people have adjusted with that particular well-water in time span and frequently they express that the water of the particular well is good in quality. The Ministry of Health, Government of India (1962),[11] has prescribed scale of physical and chemical standards of drinking water which differs from the scales given by the WHO (*Appendix-4*)

5.5 CAUSES AND EFFECTS OF POLLUTED WATER SUPPLY

Rain water running on surface dissolves many pollutants and increases pathogens which add to the pollution of ground water. If the soil-cover of the area acts as a good filter then it can reduce some harmful bacterial and guinea worm but chemical property of water will be unaffected. The chemical pollution in water increases either due to minerals present on the exposed surface or due to manure used for agricultural purposes. Most of the water-related diseases are some way or other concerned with the polluted water supply. These diseases can be distinguished as (i) the infectious water-related disease, and (ii) those related to the chemical property of water; in some way the softness and high nitrate level of water can also create some special diseases in the human body[12] .

Water quality often declines as population increases due to more competing demand for water and the type of contamination increased in the volume of human waste as well as the number of ways in which it can occur.[13]

In the rural area of Jaipur district even now the method of drawing water is traditional. People are fetching water by rope through un-cleaned bucket and individual family using there separate vessel which helps to accumulate different types of contaminating materials. If one family is infected by certain diseases (water transmitted), then it can spread easily to all other families fetching water through that particular well. Open wells accumulate not only buckets and other vessels for withdrawing the water, but also the bodies of frogs, mice, wastage and other small animals drowned in their efforts to reach the water, and also much dust and rubbish blown in by the wind.

Based on the four mechanisms by which a disease may be water-related, a recent taxonomy of water related disease has been proposed[14] and presented in table 5.3 alongwith the suggested preventive strategy to control them to some extent[15]. It is noteworthy that here only those diseases are included which are common in the Jaipur district.

5.6 WATER QUALITY CONTROL METHODS

The analysis of the condition and quality of the water supply system clearly reveals that in time span the problem of health hazard is increasing. Lack of people's participation in government scheme is one of the responsible factors in this context.

Among the various measures to check-up the bacterial growth in the polluted water, Chlorine is very popular and useful as it acts to destroy the enzymes essential for the existence of micro-organisms. Additionally, it also oxidizes iron, manganese hydrogen sulphate, along with destroying taste and colour-producing constituents; controlling algae and slime, and also aiding coagulation.[16] It is also suggested that water should not be stored in huge basins, because this will promote the growth of malarial diseases.

5.7 MAJOR PROBLEM OF FLUOROSIS IN RURAL JAIPUR DISTRICT

The silence is the first thing to hit you in the village of Jharana Khurd, 60 km from Jaipur. Then you realise why: There are no young people in the village. Every one of the 1,200 residents, no matter if they are 50, 30 or 10, looks old, with yellow, cracked teeth and pronounced limps. Blame it on fluorosis, triggered by the fluoride pumped up from the heart of the earth through the only functional tube wells. It's the story of all the 146 villages in Phagi and Chaksu tehsil, dominated by

Table 5. 3 : Classification of infective diseases in relation to water supplies

No.	Category	Disease	Mode of spread	Preventive Strategy
*1	Water Borne	Typhoid, Cholera, Infective Hepatitis	Spread through water supplies.	1. Improvement of water quality. 2. Prevent casual use of other un improved sources
*2	Water Washed	Scabies, Trachoma, Bacillary dysentery	Spread due to the lack of water for personal hygiene.	1. Improvement of water quality, 2. Make available greatest volume of water.
*3	Water Based	Schistosomiasis, Guniea Worm	Spread through an aquatic invertebrate animal.	1. Protection of water source. 2. Control of snail population.
*4	Water Related Insect Vectors	Malaria	Spread by insect that depend on water	1. Improvement of surface water management. 2. Destroy breeding sites of insects.
5.	Water Quality		Spread through water supplies.	1. Improvement of water quality. 2. Prevent casual use of other un improved sources 3. Use water quality control methods
	(1) Fluoride	Fluorosis		
	(2) Nitrate	Methemoglobinema		

(* Based on Bradley, 1974).

labourers and small farmers, a tragedy underlined by three-and-half years of drought. The failure of rains has caused groundwater levels to drop and fluoride levels to concentrate alarmingly.[17]

Jaipur district is more effected district with the problem of excessive fluoride in ground water. The village Nasnauta, Bagawas, Junsiya khurd, Baika Bas and Goner are having fluoride concentration more then 11 mg/L (Table 5.4). In Jaipur district geological and geographical conditions are such that ground waters are rich in mineral and toxic constituents like nitrates and fluorides. Surface sources are meagre and most of the supplies are dependent on the ground water sources.

There are more than 491 villages, where fluorides are more than 3 ppm and 1187 villages where fluorides are above 1.5 ppm.[18] Some villages have already been covered with protected drinking water supply through Regional water supply schemes.

5.7.1 Sources and Effects of Fluorides

Fluoride is so highly reactive that it is never encountered as a free element in nature. The possible channels through which fluoride makes a way into the human bodies are ground water, tea, fish, meat and peas. Tea and peas are very high in fluoride content depending on the fluoride content of the soil in which they are grown. Air also becomes a source of high fluorides near aluminium and other factories where fluorine or its compounds are used. Beneficial fluoride turns to a problematic parameter when present in concentrations exceeding 3 mg/l. The effects are more apparent in children as the fluoride content in their bones is very low compared to that of adults.

Table 5.4 : Places with high fluoride content (More than 1.5 mg/L in Ground Water)

No.	Location	Block	Fluoride mg/L
1.	Chaksu	Chaksu	2.00
2.	Nasnauta	Dudu	11.00
3.	Phagi	Phagi	2.90
4.	Sanganer	Sanganer	7.30
5.	Paota	Kotputli	5.00
6.	Jatwara Rly. Stn.	Bassi	2.30

Contd...

...Contd.

No.	Location	Block	Fluoride mg/L
7.	Baika Bas	Sambhar	15.60
8.	Jobner	Sambhar	1.80
9.	Dawach	Chaksu	1.80
10.	Mangarwala	Dudu	1.80
11.	Padsoli	Dudu	4.50
12.	Pallu Khurd	Dudu	6.60
13.	Thali	Chaksu	4.70
14.	Sirohi Khurd	Dudu	1.60
15.	Kalwad	Jhotwara	2.45
16.	Goner	Sanganer	11.75
17.	Hastal Ka Bas	J. Ramgarh	5.00
18.	Chardera	Kotputli	3.00
19.	Banhar	Kotputli	7.50
20.	Rakanpura	Kotputli	2.75
21.	Malpura	Kotputli	3.50
22.	Paniala	Kotputli	3.45
23.	Saraund	Kotputli	2.75
24.	Pawana	Kotputli	1.70
25.	Ajitpura	Kotputli	1.54
26.	Sujat Nagar	Kotputli	3.75
27.	Fatehpur	Kotputli	7.50
28.	Chhatariwali Dhani	Kotputli	4.70
29.	Astal	Kotputli	1.60
30.	Mandha	Kotputli	2.00
31.	Dhani Nakkai	Kotputli	2.00
32.	Nangal	Kotputli	2.75
33.	Kansli	Kotputli	2.00
34.	Sundarpura	Kotputli	9.00

Contd...

...Contd.

No.	Location	Block	Fluoride mg/L
35.	Tapri	Kotputli	7.30
36.	Malahera	Kotputli	2.25
37.	Hastera (EW)	Govindgarh	3.70
38.	Bhojpura (EW)	Sambhar	2.10
39.	Bagawas (EW)	Sambhar	1.80
40.	Bassi (EW)	Bassi	2.40
41.	Pilwa	Amber	3.60
42.	Champapura	Jhotwara	4.00
43.	Malwara	Jhotwara	3.25
44.	Palri	Jhotwara	7.30
45.	Mohana	Sanganer	8.20
46.	Tilwara	Sanganer	4.50
47.	Jaipur	Jaipur	2.24
48.	Gujran Ki Talai	Sanganer	5.00
49.	Kanarwas	Bassi	8.00
50.	Kanota	Bassi	3.10
51.	Watika	Sanganer	3.50
52.	Ramsinghpur	Sanganer	3.60
53.	Dhaurala	Sanganer	4.80
54.	Manglya	Sanganer	5.60
55.	Mohanpura	Bassi	3.00
56.	Damodarpur	Bassi	3.75
57.	Mundahar	Bassi	6.80
58.	Kanauta	Bassi	3.10
59.	Bagru	Sanganer	2.00
60.	Samhori	Jhotwara	1.60
61.	Shyasinghpura	Sambhar	4.50
62.	Sawatka Bas	Sambhar	5.80

Contd...

...Contd.

No.	Location	Block	Fluoride mg/L
63.	Bagawas	Sambhar	11.70
64.	Lalasar	Sambhar	6.80
65.	Ratawata	Sambhar	2.50
66.	Junsiya Khurd	Sambhar	11.30
67.	Basdi Khurd	Sambhar	2.10
68.	Etawa	Sambhar	2.60
69.	Kanwarpure	Sambhar	3.10
70.	Dungri Kalan	Sambhar	2.00
71.	Harsoli	Sambhar	2.30
72.	Karansar	Sambhar	2.70
73.	Khijuriya	Sambhar	9.20
74.	Mohanpura	Sambhar	2.2
75.	Basdi Kalan	Sambhar	3.20
76.	Khedi Malik	Sambhar	10.00
77.	Mohan Ka Bas	Sambhar	8.80
78.	Dhankawas	Sambhar	4.90
79.	Kishangarh Renwal	Sambhar	7.20
80.	Sukhalpura	Sambhar	2.40
81.	Bhojpura	Sambhar	2.30
82.	Kodi	Sambhar	4.80
83.	Pachkodiya	Sambhar	6.60
84.	Jobner	Sambhar	6.60
85.	Dhanikalan	Sambhar	2.80
86.	Mohana	Sanganer	2.05

Source : PHED, Govt. of Rajasthan, Jaipur.

In addition to dental and skeletal fluorosis, it causes decreased haemoglobin and increased erythrocyte levels in children. Individuals with pulmonary and renal disfunction are easily susceptible to fluoride intoxication. When human beings are exposed to high fluoride concentration for longer periods, it causes depression in collagen formation, bone resorption, increase in bone crystal and decrease in aortic calcification.

5.7.2 Methods of De-Fluoridation (DF)

The methods of removal of undesirable concentration of fluorides consist of either chemical precipitation, settling and filtration or process of ion exchange. The following are the various Defluoridation methods practised in the area.

1. Ion Exchange Method
2. Nalgonda Technique
   (a) Domestic Defluoridation
   (b) Fill & Draw DF system for rural water supply.

5.7.2.1 Ion Exchange

Ion Exchange is one of the methods used for the removal of undesirable concentration of fluorides. Usually fluoride concentration encountered in raw water is small ranging between 5 to 10 PPM and is required to bring down to 1.0 to 1.5 PPM. Ion exchange process is found suitable and economically viable. Fluoride bearing water is passed over a bed of suitable granular material which reacts with it chemically and adsorbs the resulting product. With the passage of time when bed gets exhausted, its exchange capacity is replenished by passing suitable reagents through it. Thus it can be reused for several times. The materials commonly used for Ion-exchange process are Calcium phosphate, Fluorex, Lime, Activated carbon and Patented materials like Defluoron -2.

5.7.2.1 Nalgonda Technique

The National Environmental Engineering Research Institute (NEERI), Nagpur designed and erected defluoridation plant at Nalgonda in Andhra Pradesh for local municipality. The method for the removal of excessive fluoride at domestic, community and rural water supply levels using precipitation by aluminium salts is known as Nalgonda Technique (Figure 5.5).

The sequence of treatment is precipitation, settling and filtration. This involves addition of aluminium salt, lime and bleaching power followed by rapid mixing, flocculation, sedimentation, filtration and disaffection. Aluminium salt may be added as aluminium sulphate

Defluoridation Plant Attached with Handpump Nalgonda Technique.

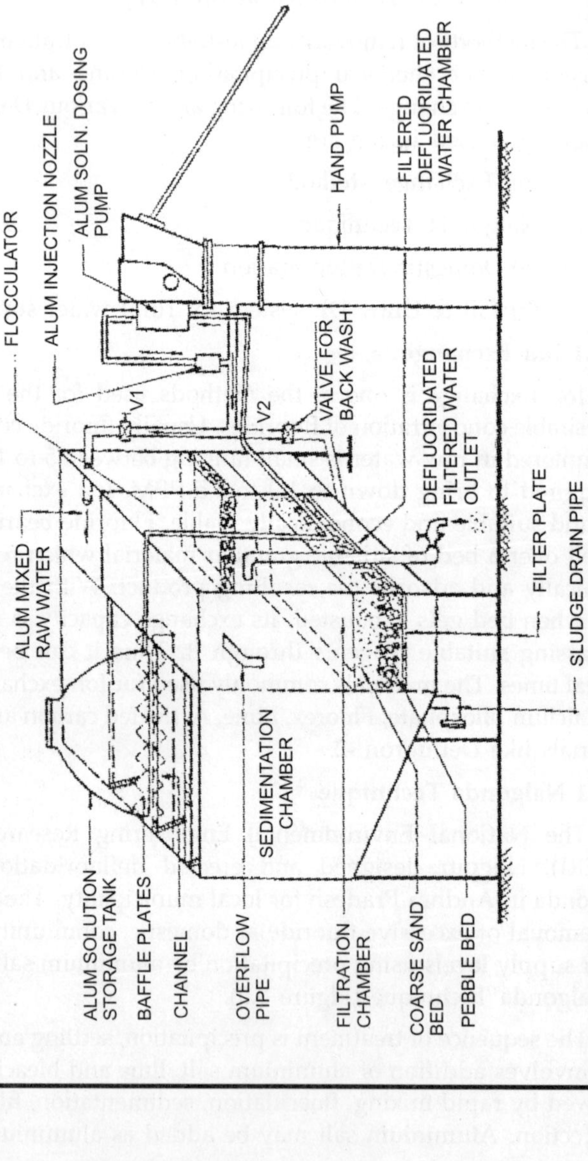

Figure : 5.5

* Source : NEERI, Jaipur.

or aluminium chloride or combination of these two Aluminium salt is effective for removal of fluorides from water.

The dose of aluminium salt increases with the increase in fluoride and alkalinity levels of the raw water, the selection of either aluminium sulphate or aluminium chlorides also depends on sulphate and chloride content of raw water to avoid exceeding their permissible limits. The dose of lime is empirically 1/20th that of the dose of aluminium salt. Lime facilitates forming denser floe for rapid settling. Bleaching power is added to the raw water at the rate of 3 mg/l for disaffection. Approximate doses of alum required to obtain permissive limit (1 mg/l) of fluoride in water at various fluoride and alkalinity levels are given in Table 5.5.

Table 5.5 : Approximate Alum Dose (mg/l) Required to Obtain Permissive Limit (1 mg/l) of Fluoride In Water At Various Alkalinity and Fluoride Levels

Test water	Test Water Alkalinity mg Ca CO_3/L							
Fluoride mg/l	125	200	300	400	500	600	800	1000
2	143	221	273	312	351	403	468	520
3	221	299	351	403	507	520	585	767
4	•	403	416	468	559	598	689	936
5	•	•	507	598	689	715	884	1010
6	•	•	611	715	780	936	1066	1209
8	•	•	•	•	988	1118	1300	1430
10	•	•	•	•	•	•	1508	1690

• Can be treated only after rising alkalinity with either lime or sodium carbonate.

Source : PHED, Govt. of Rajasthan, Jaipur.

The presence of calcium and magnesium hardness, sodium, potassium, sulphate, chloride, nitrate, phosphate, organic matter and residual chlorine in water do not have any adverse effect on removal of fluoride by this method. The chemical reaction involving fluorides and aluminium is complex. It is a combination of polyhydroxy aluminium species complexation with Fluorides and their adsorption on polymeric aluminium hydroxide, flocs. Besides fluorides, turbidity, colour, odour, pesticides and organic materials are also removed and the bacterial load is also reduced significantly. All these are reduced by adsorption on flocs.

Alkalinity, natural or supplemented by the addition of lime, sodium carbonate or bicarbonate ensures effective hydrolysis of aluminium salts leaving no residual aluminium in treated water. The chemical reactions which depend on the nature of alkalinity are :

(i) $3\ Al_2(SO_4)_3\ 18H_2O + NaF + 9\ Na_2CO_3 - [5Al\ (OH)_3\ Al\ (OH)_2F] + 9\ Na_2SO_4 + NaHCO_3 + 8CO_2 + 45\ H_20$

(ii) $3\ Al_2(SO_4)_3\ 18H_2O + NaF + 17\ Na_2CO_3 - [5Al\ (OH)_2\ Al\ (OH)_2F] + 9\ NaSO_4 + 17CO_2 + 54\ H_2O$

Sedimentation following chemical mixing, permits the resultant flocs loaded with the fluorides, turbidity, bacterial and other impurities to settle thus reducing the concentration of suspended solids. The turbidity of the supernatant water, if not well within the limits, it is filtered through a rapid sand filter. The filtered water is collected in the storage tanks and is re-chlorinated if necessary with bleaching powder.

Advantage of Nalgonda Technique

(1) economical and easily adaptable method by rural population.

(2) it can be used for a wide range of flow i.e. 19 m^3/day to 480 m^3/day.

(3) simple with respect to operation and maintenance.

Nalgonda Technique Application

Based on the volume of water to be treated this technique can be applied at following levels.

(1) Domestic for individual houses.

(2) Fill and Draw type for Rural Water Supply Schemes.

(3) Continuous flow type for larger volumes.

5.7.2.1.1 ***Domestic Defluoridation Plants***

Defluoridation at domestic level can be achieved in a container (bucket) of 60 litre capacity fitted with a tap 3-5 cm. above the bottom of the container for withdrawal of treated water after precipitation and settling. The raw water taken in the container is mixed with adequate amount of aluminium sulphate solution (alum), lime and bleaching powder depending upon its alkalinity and fluoride contents. Alum

Fill and Draw Type Defluoridation System for Rural Water Supply.

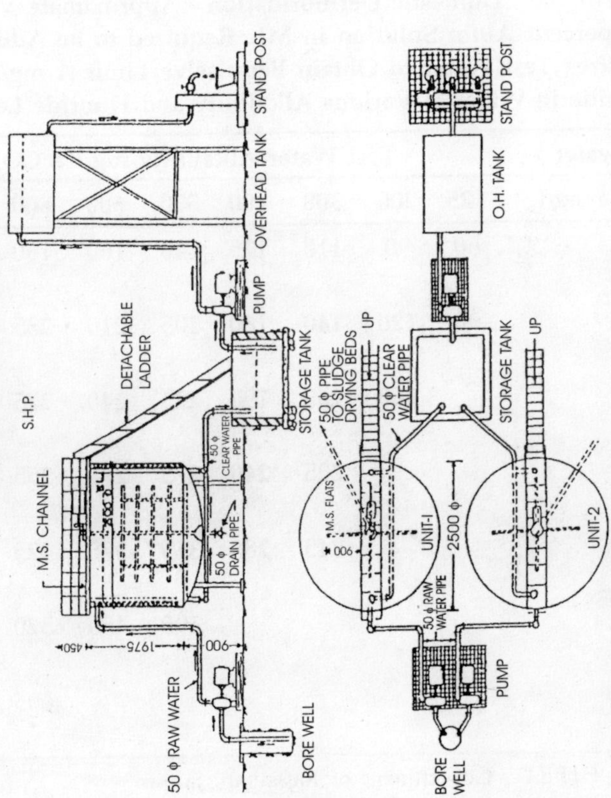

Figure : 5.6

* Source : NEERI, Jaipur.

is added first and mixed well with water. Lime slurry is then added, the contents are stirred slowly for 20 minutes and allowed to settle for nearly one hour. The supernatant water which contains permissible amount of fluoride, is withdrawn through the tap for consumption. The settled sludge is discarded. Approximate volumes of 10 percent alum solution for defluoridation of 40 litre of water are given in Table No. 5.6.

Table No. 5.6 : Domestic Defluoridation - Approximate Volume of 10 percent Alum Solution in ML. Required to be Added in 40 Litres Test Water to Obtain Permissive Limit (1 mg/l) of Fluoride in Water at Various Alkalinity and Fluoride Levels

Test water	Test Water Alkalinity mg Ca CO_3/L							
Fluoride mg/l	125	200	300	400	500	600	800	1000
2	60	90	110	125	140	160	190	210
3	90	120	140	160	205	210	235	310
4		160	165	190	225	240	275	375
5			205	240	275	290	355	405
6			245	285	315	375	425	485
8					395	450	520	570
10							605	675

Source : P.H.E.D., Government of Rajasthan, Jaipur.

5.7.2.1.2 *Fill and Draw Type Defluoridation Plants*

Fill and Draw type technique is used by PHED department in Shivdaspura, Chaksu Block. The scheme includes reaction-cum-sedimentation tanks, power driven agitator, sump well and elevated service reservoir as shown in figure 5.6. Raw water is pumped into the reactors and the required amounts of bleaching powder, lime and alum are added. The contents are stirred for 20 min. and allowed to settle for two hours. The supernatant water is collected in sump well, chlorinated and pumped to an elevated service reservoir from where the water is supplied by gravity. The settled sludge is withdrawn every day and dried on sludge drying beds.

REFERENCES

1. WHO (1971) : *International Standards for Drinking Water,* 3rd. edition, Geneva, WHO.
2. Pesod, M.B. (1977). "Surface Water Quality Criteria for Tropical Development Countries", In : *Water Wastes and Health in Hot Climate,* Feachem, R. *et. al.* eds., John Wiley, London, p.56.
3. Dixey, F. (1950) : *A practical Hand Book of Water Supply.* Thomas Murphy & Co., London, p. 304.
4. Folwell, A.P. (1917) : *Water Supply Engineering,* 3rd ed. John Wiley, New York, pp. 23-52.
5. Goel, P.K. & Sharma, K.P. (1996) : *Environmental Guidelines and Standards in India,* Technoscience Publication, Jaipur. p. 16.
6. Ministry of Health, Government of India (1962): *Public Health Engineering Manual and Code of Practices,* Section 1-As *Manual on Water Supply,* Government of India, New Delhi.
7. Washing, P H. (1949): "Significance of Nitrate in Water Supplies", *Am. Water Work Assoc. J1.,* Vol. 41, no. 2, pp. 147-150.
8. Hem, J.D. (1970). Study and Interpretation of the Chemical Characteristics of Natural Water, 2nd ed. U.S. Geol. Survey Water Supply Paper-1473, p. 363.
9. Lohr, E.W. and Love, S.K. (1954): *The Industrial Utility of Public Water Supplies in U.S.,* pt. 1, U.S.Survey Water Supply Paper- 1299, p. 639.
10. U.N. Water Resource (1976): *Demand for Water, Procedures and Methodologies for Projecting Demand in the.Context of Regional and National Planning,* ST/ESA/38.
11. Ministry of Health, Government of India (1962) : *ibid* (ref. 6)
12. Feachem, R. (1975) : "The Rational Allocation of Water Resources for the Domestic Need of Rural Communities in Developing Countries" in *Proceedings of Second World Congress of Water Resource,* New Delhi, Vol.2 p.539-547.
13. Burton, I. (1974): "Domestic Water Supplies for Rural People in the Developing Countries: The Hope of Technology", In *Human Rights in Health,* Eilliott, K. and Knight, J. eds., Ciba Foundation Symposium, 23 (New Series) Amsterdam, ASP -North Holland, pp. 61-79.

14. Bradley, D.J. (1974): "Water Supply: The Consequence of Change" (Elliott, K. and as Knight, J. eds.), In: *Human Right in Health*, Amsterdam : ASP North-Holland, pp. 81-98.

15. Feachem, R. (1975) : "Water Supplies for Low-Income Communities in Developing Countries", *Journal of the Envi. Eng. Division*, ASE, Vol. 101, p. 687-702.

16. CPHERI (1972) : *Disinfections of Well Water (Technical Digest no. 26*, Nagpur, Central Public Health Engg. Res. Institute: also Pickford, J. (1977): "Water Treatment in Developing Countries", In *Water and Health in Hot Climate*, Feachem, R. *et. al.* eds., John Wiley, London, p. 186.

17. Website : *http://www.fluoridealert.org/news/india/rajasthan.htm*, web site of Fluoride Action Networks.

18. PHED (1998) : *An Overview : Drinking Water*, Public Health Engineering Department, p. 8

□□□

6

Rural Domestic Water Supply

Water is the basic necessity of human life. But as a matter of fact only an insignificant portion (2 percent) of the entire water mass available in the hydrosphere is potable[1]. Without an ample supply of potable water, all the body functions suffer and the body degenerates, because air alone cannot clean the blood sufficiently and food is imperfectly assimilated[2]. Man has over 60 percent of water by weight, constituting 26.5 litres inside the cells and 3.5 litres in blood[3]. This being one of the basic necessities of life, Government has come out with various programmes for the supply of drinking water. The piped water supply is an economical proposition for areas with high population density. But the problem of supply of safe drinking water assumes serious proportion when the efforts relate to the large rural population scattered over a vast area including isolated and inaccessible habitations[4].

In ancient times, towns and villages were invariably situated on or near water bodies, preferably the banks of streams from which they conveniently got water but in time span when population increased and extended inside the country, lacking natural supplies, artificial methods of securing the requisite supply became necessary, consequently there developed different water drawing methods.

6.1 WATER SUPPLY SCHEMES

Sources of water supply depend almost wholly upon the rainfall and resulting water bodies like springs, rivers and lakes. Water bodies

Table 6.1 : Water Supply Schemes in Jaipur District (2001)

Numbers (Percentage)

No	Tehsil	No. of Villages			Water Supply Schemes					
		Total	Unpopulated	Populated	No any Scheme	Piped Scheme	P&T Scheme	Regional Scheme	T.S.S	Hand Pump Scheme
1	2	3	4	5	6	7	8	9	10	11
1	Phulera	215	0	215	0	15	12	70	0	118
				(100)	(0.00)	(6.98)	(5.58)	(32.56)	(0.00)	(54.88)
2	Phagi	170	2	168	0	5	7	26	0	130
				(100)	(0.00)	(2.98)	(4.17)	(15.48)	(0.00)	(77.38)
3	Sanganer	209	4	205	0	6	2	26	0	171
				(100)	(0.00)	(2.93)	(0.98)	(12.68)	(0.00)	(83.41)
4	Chaksu	279	21	258	5	4	6	94	1	149
				(100)	(1.94)	(1.55)	(2.33)	(36.43)	(0.39)	(57.75)
5	Dudu	147	9	138	0	12	3	29	2	92
				(100)	(0.00)	(8.70)	(2.17)	(21.01)	(1.45)	(66.67)
6	Amber	196	2	194	1	8	20	6	4	154
				(100)	(0.52)	(4.12)	(10.31)	(3.09)	(2.06)	(79.38)

Contd...

...Contd.

1	2	3	4	5	6	7	8	9	10	11
7	Bassi	214	5	209	0	6	14	14	1	174
				(100)	(0.00)	(2.87)	(6.70)	(6.70)	(0.48)	(83.25)
8	Chomu	95	0	95	0	24	17	4	0	50
				(100)	(0.00)	(25.26)	(17.89)	(4.21)	(0.00)	(52.63)
9	J. Ramgarh	241	10	231	0	8	9	9	9	196
				(100)	(0.00)	(3.46)	(3.90)	(3.90)	(3.90)	(84.85)
10	Kotputli	137	0	137	0	13	16	8	0	100
				(100)	(0.00)	(9.49)	(11.68)	(5.84)	(0.00)	(72.99)
11	Shahpura	89	0	89	0	11	6	0	13	59
				(100)	(0.00)	(12.36)	(6.74)	(0.00)	(14.61)	(66.29)
12	Viratnagar	93	3	90	0	6	9	0	3	72
				(100)	(0.00)	(6.67)	(10.00)	(0.00)	(3.33)	(80.00)
13	Jaipur	102	0	102	0	4	11	0	13	74
				(100)	(0.00)	(3.92)	(10.78)	(0.00)	(12.75)	(72.55)
Total				2131	6	122	132	286	46	1539
				(100)	(0.28)	(5.73)	(6.19)	(13.42)	(2.16)	(72.22)

Source : Public Health Engineering Department, Jaipur.

are influenced by the geological structure which enables certain rocks to store that part of the rain which percolates through the soil. Except the few area of Jamwa Ramgarh, Viratnagar tehsil and northern part of the district, where people are utilising the surface water, dug wells and tube-wells (underground water) are common in all parts of the Jaipur district. These wells are seldom properly protected from underground and surface contamination; however, they can be considered as excellent sources if the water table is high and harmful minerals are absent in the aquifers.

Before looking into the future needs of drinking water for Jaipur district, it is necessary to make a correct assessment of the existing coverage of water supply schemes in the Jaipur district. For planning of the different schemes, as already explained, a long term perspective has to be kept in view[5]. There are 2131 populated villages out of 2187 villages in Jaipur district and 2126 villages have been covered with drinking water supply with PHED (Public Health Engineering Department) in year 2000-2001 (Table 6.1). Five main type of schemes are being executed by PHED for water supply in rural areas.

6.1.1 Pump and Tank Schemes (P&T)

In Jaipur district, 6.19 percent villages were under this scheme in 2001 (Table 6.1). The Pump and Tank Scheme is being run by the Public Health Engineering Department (PHED). The villages with population between 2000 to 5000 inhabitants are covered under these schemes. Among these schemes the more popularly known scheme is *Panghat Schemes*. Under this scheme the source of water, well/borewell is located in the same village[6].

This scheme is further linked with Pipe scheme, Regional scheme and Public Stand Post (PSP). The scheme has a ground level service reservoirs, constructed at a central location in the village. Public stand posts are provided around the water supply for the villagers to fetch water from the PSP and carry it to their houses.

In the P & T scheme water is drawn from different sources and is collected in a large clear water reservoir tanks. Mostly three types of sources are utilised for drawn water.

6.1.1.1 *Underground Water Source*

The P & T scheme mainly utilises the underground water source. Underground water is pumped through electric motors from various place and collected in a clear water reservoir tanks via a main water line which is called rising line. The diameter of the rising line increases with each connection to different sources, finally reaching the clear water reservoir.

6.1.1.2 *Surface Water*

At some places (Jamwa Ramgarh tehsil) water from surface water source like lake, dams, rivers etc. are taken through suction pumps and this water is passed through the filter plant to remove the contained impurities like sand, natural products etc. This filtered water is stored in clear water reservoir through rising main. This technique of filtration is not adopted in underground water, as the underground water does not contain this type of impurities.

6.1.1.3 *Traditional Water Source*

Water from traditional sources like ponds, wells etc. is taken through electric motor pumps and is filtered by passing through the filter plants. The filtered water is stored in clear water reservoir tanks by passing through rising main. The filter plants are attached in the water line which are anticipated to contain substantial amount of removable impurities.

Water collected from all these type of sources is stored in Clear Water Reservoir (CWR) tank which is a huge underground artificial tank, whose capacity depends on the population of the served area and connected sources of electric motor pump. The water inflows in the rising main and is stored in the CWR tank for 24 hours a day. Chlorine gas is passed in CWR tanks to clear harmful bacteria and viruses form stored water. The water of CWR tanks is lifted through motor pump to the Service Reservoir (SR). SR tank is a special type of tank built 90 to 100 feet above the ground level and its height is also above the height of the served area so that enough pressure is maintained during supply. A motor pump is connected between SR tank and CWR tank. An special pipe called suction pipe connects the motor pump to the CWR tank and another pipe called delivery pipe connects the motor pump to the SR tank. Some different pipes are connected to the SR tank for different purposes.

- **Rising Pipe**

  With the help of this pipe water is collected in the SR tank through pump. One end of this pipe opens in SR tank and another is connected to the delivery pipe.

- **Wash out pipe**

  This is a special type of pipe one end of which is connected to the bottom of the SR tank and a stop cork is inserted into another end of this pipe. This pipe is used during the cleaning of the SR tank.

- **Overflow Pipe**

  This pipe flows from the top of SR tank to the main distribution line below. The water is passed into the distribution channels directly to save the water during overflow.

- **Distribution main line**

  Water is supplied through distribution main line (Distribution main line is further divided into sub-distribution lines and these are further ramified into distribution line). These distribution lines flow the water to the houses. A water meter is connected across to every distribution line in front of the consumer house. This meter records the quantity of water consumed and PHED gives bills to the consumers accordingly.

In Jaipur district 132 villages use the P & T schemes (Table 6.1). The percentage of this scheme area is not uniformly equal in every tehsil of the Jaipur district. High percentage of village under P & T scheme are concentrated in northern part of the district, like 17.89 percent in Chomu, 11.68 percent in Kotputli, 10.31 percent in Amber and 10.00 percent in Viratnagar (Figure 6.1). Low and very low percentage of village are using this scheme in southern part of district, like Sanganer (0.98), Dudu (2.17 percent) and Chaksu tehsil (2.33 percent) (Figure 6.1).

During the period of 1991 to 2001 the number of villages under this scheme increased from 104 to 132 villages. This rate of increase is not uniformly equal through the district. The maximum number of villages under this scheme increased in all of the tehsils except Chomu, Shahpura, Viratnagar and Phulera tehsils.

6.1.2 Piped Scheme

Pipe scheme is also run and executed by the PHED. The villages having population of more than 5000 inhabitants are covered under this scheme[7]. Under this Scheme water is supplied through the pipes and against which PHED bills the consumer accordingly. They also provide for house connections, public stand posts and cattle water troughs.

SR tank is the main source in the pipe scheme, through which main distribution line is connected. At some places CWR tank is used directly for the supply instead of the SR tank. By this scheme the area of Viratnagar, Kotputli, Jaipur, Amber, Chomu and Sanganer tehsils are served twice a day and other areas once a day[8]. In this scheme main distribution line is connected to the sub-distribution line of smaller diameter and these are ramified into small distribution line of lesser diameter. These distribution lines are connected to the households via water meters.

In Jaipur district 122 villages use the Piped schemes (Table 6.1). High percentage of village under Pipe scheme are concentrated in northern-western part of the district, i.e., Chomu (25.26 percent), Shahpura (12.36 percent) and Kotputli (9.49). Very low percentage of village are using this scheme in southern-eastern part of the district, i.e. Chaksu (1.55 percent), Bassi (2.87 percent) and Phagi tehsil (2.98 percent) (Figure 6.1).

During one decade (1991 to 2001) the number of villages under Pipe scheme increased from 72 to 122 villages. This rate of increase is not uniformly equal through the district. The maximum number of villages under Pipe scheme increased in all the tehsils except Dudu tehsil, which has low potentiality of ground water and scattered patches of brackish water.

6.1.3 Regional Schemes

Majority of population in village resides in settlements away from main village as a result of which the population is not fully benefited though village has been provided with drinking water supply. Such villages are mainly in uncultivated or desert area where coverage is through regional water supply schemes, where only one water supply point such as GLR or PSP is provided [9]. Villages which do not have a potable source within the boundary are supplied through nearby potential source by a pipeline. Such type of schemes are called regional schemes.

Regional schemes are run by PHED and *Gram Panchayats* collectively. The installation and implementation of this scheme is done by PHED and the execution and maintenance is done by the

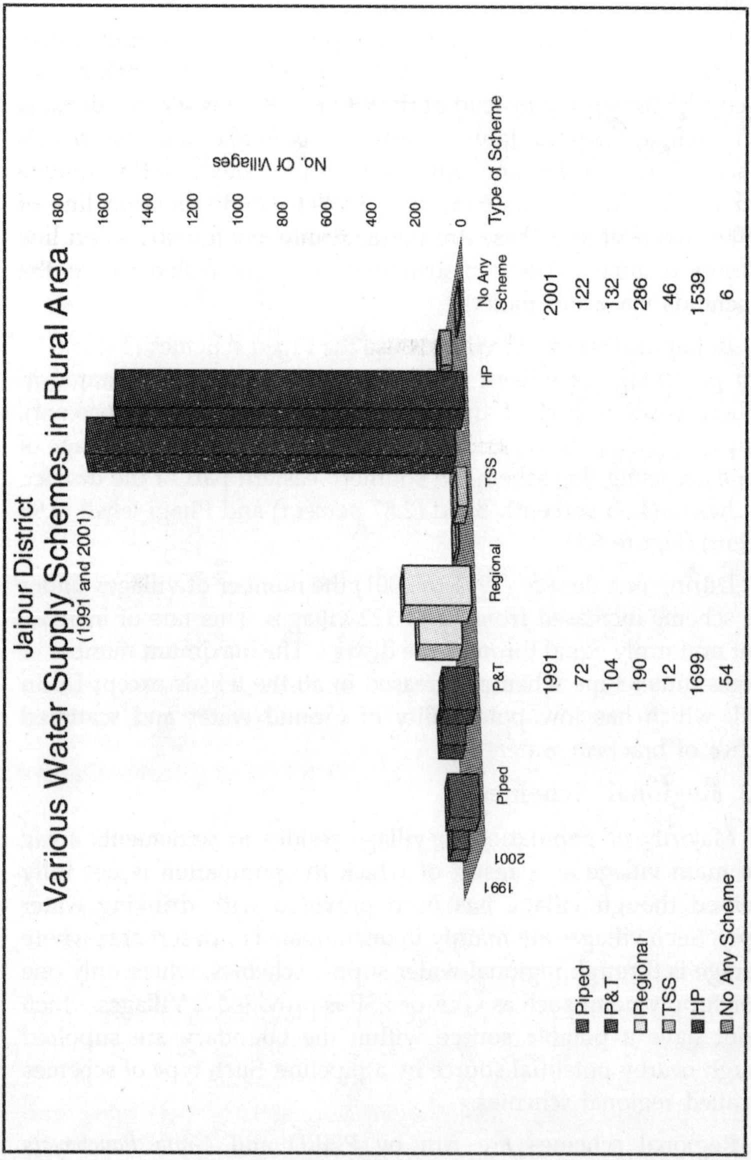
Jaipur District
Various Water Supply Schemes in Rural Area
(1991 and 2001)
No. Of Villages
1800
1600
1400
1200
1000
800
600
400
200
0
1991
2001
Piped
P&T
Regional
TSS
HP
No Any Scheme
Type of Scheme
1991 72 104 190 12 1699 54
2001 122 132 286 46 1539 6
Piped
P&T
Regional
TSS
HP
No Any Scheme

Figure : 6.1

Gram Panchayats. This scheme is implemented by combing five to six villages (combination of few villages called *"Dhani"*) together. For smooth working of this scheme PHED provides the electricity expenses of electric motor pump @ Rs. 400/- per HP (House Power) and repairing expenses of the motor and Rs. 500/- against wages of one person who look after the fitting and maintenance of tank. Other remaining expenses and complete maintenance are borne by the *Gram Panchayats*.

Under this scheme a Ground Level Reservoir (GLR) is established in every village. The shape, size and capacity are decided as per the population of the village. GLR tanks of every village is connected to the source of water by a main line. The main water source is established among one of these villages. This source is connected through pump and tank scheme to the CWR. With the help of electric motor, water from this source is taken upto the GLR tanks. A person is appointed by *Gram Panchayats* for looking after and filling of the GLR tanks in every village. PHED provides Rs. 500 per month to the *Gram Panchayats* against the wages of this person.

GLR tank is built above the ground level and its floor is parallel to the ground. Many pipes come out of the GLR tanks and taps are connected across the pipes through which the consumer use the water. In this scheme there is no other charge of water for consumer.

286 villages of Jaipur district use Regional schemes. Maximum percentage of village under Regional scheme are concentrated in south-western part of the district, i.e., Chaksu (36.43 percent), Phulera (32.56 percent), Dudu (21.01 percent) and Phagi (15.48 percent), where such villages are mainly in uncultivated or desert area. Regional scheme is not used in three tehsils (Shahpura, Viratnagar and Jaipur tehsil) of Jaipur district, where main problem is of industrialisation in rural area. Low percentage of village are using this scheme in middle and eastern part of district.

During the period of 1991 to 2001 the number of villages under Regional scheme increased from 190 to 286 villages (Figure 6.1). The maximum number of villages under this scheme increased in south-western part of district, i.e. Phulera (form 39 to 70 villages) and Chaksu tehsil (form 39 to 94 villages), where main problem is of high fluoride content in underground water. Regional scheme have decreased in Shahpura (form 8 to nil) and Viratnagar (form 1 to nil), because of the impact of water pollution and industrialisation in this area.

6.1.4 Traditional Source Scheme (TSS)

In Jaipur district only 2.16 percent villages were under this scheme in 2001 (Table 6.1). Traditional Source Scheme (TSS) are run through *Gram Panchayats*, municipalities and PHED collectively. Under this scheme renovation and maintenance of old traditional sources, water purity maintenance and rain water collection is managed by *Gram Panchayats* and various village development committees. *Gram Panchayats* and village committees pay focus on rain water collection in these traditional sources, apart from regular maintenance. This help in increasing the water level continually of the source.

In Jaipur district only 46 villages is using Traditional Source scheme (Table 6.1) and it is implemented in only 8 tehsils out of the 13 tehsils. High percentage of village under TSS are concentrated in Shahpura (14.61 percent) and Jaipur tehsil (12.75 percent). TSS is not used in five tehsils (Phulera, Phagi, Sanganer, Chomu and Kotputli tehsil) of Jaipur district. During one decade (1991 to 2001) the number of villages under TSS increased from 12 to 46 villages (Figure 6.1).

6.1.5 Handpumps

First of all it has to be recognised that handpump is a "*Mini Water Works Unit*", by itself, combining in a simple, but effective manner, all the complicated processes for the exploitation of water source, natural treatment, simple pumping, distribution and the last but not the least even the effective disposal of waste water generated[10].

This scheme is extensively used in rural areas. The smaller villages having population of less than 2000 inhabitants are mainly benefited from these schemes. One handpump is provided for every 250 inhabitants. After making, provision for standby hand pumps needed for repairs, a norm of one handpump for 150 population is considered adequate. At any time the number of working hand pumps should be one for every 250 population. The water is drawn by the consumer directly from the handpump and he carries it to the house. PHED, *Gram Panchayats*, municipalities and regional committees are responsible for installation, execution and maintenance of these handpumps. Generally, handpumps are installed at a place where underground water is clear and portable and is available upto the depth of 200 feet. In this scheme there is no any charges for installation of handpump. The guiding objective is that the handpump should provide trouble free service to all the users especially during the critical period of summer. It should provide drinking water to the people when their needs for water is paramount. The norms should be geared to achieve this basic and fundamental objectives while at the same time ensuring *"Cost Effective"* maintenance.

Table 6.1 shows that highest coverage (72.22 percent) of water supply schemes is through handpumps in Jaipur district. The maximum coverage (above 80 percent) of handpumps schemes is in the eastern parts of the district due to peneplain, high irrigation intensity and also more ground water potentiality. Medium coverage (60-80 percent) of the handpumps schemes is in the middle part and northern part of the district. Low and very low (less than 60 percent) areas under handpumps schemes are found in the southern and western parts of the Jaipur district, where low ground water potentiality and annual average rainfall is lowest.

During the period of 1991 to 2001 the number of villages under Handpump scheme decreased from 1699 to 1539 villages (Figure 6.1). The maximum number of villages under this scheme decreased in Chaksu tehsil (form 207 to 149 villages), where main problem is of high fluoride content in underground water and have low potentiality of ground water.

In respect of regional schemes the quality of water being supplied to individual/component villages shows that the quality of water has improved and the village seizes to remain problem village with regard to its chemical parameters. Generally supply of water in the schemes covered by PHED is as per standards applicable for Rajasthan. In case of handpumps it has not been possible to check the quality parameters of the water being supplied.

6.1.5.1 *Handpump Maintenance System*

Handpumps are basically mechanical appliances, requiring regular maintenance and care to avoid breakdown. Unfortunately in Jaipur district handpump have decreased and numbers of abandoned handpump have increased rapidly during last eight years (1993 to 2001), because it is used by all sections of the rural community and as such it is subject to all sorts of handling, misuse and even abuse.

Frequent breakdown of handpump has forced the villagers to go back to the old unreliable water sources with all the handicaps. All the capital investment in that case has gone to waste.

The only answer to this is to ensure proper and effective maintenance of these handpumps on a systematic and regular basis. These handpumps are located at very remote places all over the district and have to work in different conditions prevalent at different places.

Table 6.2 shows, tehsil-wise, the present coverage by handpump schemes and the additional handpumps to be installed as per the norms prescribed above. A review carried out on the coverage of the handpumps indicated that out of 1539 villages where handpumps have been provided, a total of 1046 villages have inadequate number of handpumps. Therefore, it is necessary to provide additional 2623 handpumps in these villages to avoid crowding (Table 6.2). The calculation for working out adequate number of handpumps in a village a norm of one handpump per 150 persons has been adopted.

Table 6.2 : Number of Villages in Each Tehsil Having Inadequate Handpump Facilities (In handpump water supply schemes) in Jaipur District (2000-2001)

No.	Tehsil	Total No. of villages	No. of villages with Handpump Schemes	No. of villages with inadequate Handpumps	No. of additional Handpump to be installed
1	Phulera	215	118	82	228
2	Phagi	168	130	77	172
3	Sanganer	205	171	83	192
4	Chaksu	258	149	97	157
5	Dudu	138	92	58	151
6	Amber	194	154	119	215
7	Bassi	209	174	86	115
8	Chomu	95	50	38	125
9	J. Ramgarh	231	196	136	314
10	Kotputli	137	100	92	407
11	Shahpura	89	59	51	191
12	Viratnagar	90	72	59	157
13	Jaipur	102	74	68	199
Total		2131	1539	1046	2623

Source : Public Health Engineering Department, Jaipur.

Different approaches have been made with the aim of keeping large number of handpumps in working order and these fall broadly into the following categories[11]:

(i) Govt. organised maintenance.

(ii) Community or local body maintenance.

(iii) Varying combination of these types.

Most handpump maintenance systems can be characterised as one-tier, two-tier or three-tier systems. In one-tier system all maintenance tasks are the responsibility of a Government organisation, but in this system only 9 percent abandoned handpumps are repaired by PHED in year 2000-2001. In the three-tier system, the intermediate level of Government assumes specific part of the maintenance duties. In the two-tier and three-tier systems, the local community or a resident working on a voluntary basis or on remuneration assumes responsibility for basic village level handpump caretaking, maintenance and failure reporting. Routine maintenance work includes lubrication, bolt tightening and minor repairs of the handpump as well as replacement of leather washers etc. It is necessary for a technical support team to visit the pump at regular intervals for a thorough servicing of the handpump, where villagers deal with only the basic maintenance tasks above the ground level.

A handpump programme without a maintenance structure is incomplete. Maintenance is a tangible requirement and the question should be how it should be structured. The central question is whether it should be the responsibility of the Government or community, or both.

6.2 DUG-WELLS

Dug wells are common in all parts of the rural area. These wells are seldom properly protected from underground and surface contamination; however, they can be considered as excellent sources if the water table is high and harmful minerals are absent in the aquifers. Dug-wells or Drinking-wells are common since long, generally ranging from 10 m to 20 m deep and 2 m to 4 m in diameter.

People mostly fetch water with the help of rope and bucket in 5 to 20 m deep wells, therefore, it requires more manual labour. Although people do it within their daily routine, they certainly feel it as a very exhaustive work. Proper drawing arrangement on wells save manual labour in comparison to direct pulling of water buckets.

Jaipur district has 158834 dug-wells in 2000-2001, out of these wells 83.44 percent wells are useful (Table 6.3). In the district 16.56 percent (Table 6.3) wells are abandoned due to abnormal lowering of ground water level and pollution like cesspools, surface drains and refuse dumps. The maximum 13.96 percent of dug wells is in Phulera tehsil because this area is peneplain and also has more ground water potentiality. Lowest number of dug wells is in Viratnagar tehsil (2.90 percent) due to most of the hard rock strewn area. During the period of 1996 to 2000 the number of dug wells has increased from 124997 to 158834.

Table 6.3 : Tehsil-wise Wells in Jaipur District (2000-2001)

No	Tehsil	Useful Dug Wells		Abandoned Dug Wells		Total
		No.	Percent	No.	Percent	
1	Jaipur	7348	84.63	1335	15.37	8683
2	Bassi	11824	87.18	1739	12.82	13563
3	Chaksu	10426	77.68	2996	22.32	13422
4	Sanganer	12028	79.45	3111	20.55	15139
5	Amber	12963	85.90	2127	14.10	15090
6	J. Ramgarh	10144	76.88	3050	23.12	13194
7	Chomu	12587	91.63	1150	8.37	13737
8	Phulera	18195	82.03	3985	17.97	22180
9	Dudu	8144	78.81	2190	21.19	10334
10	Phagi	8978	75.82	2863	24.18	11841
11	Kotputli	9782	93.50	680	6.50	10462
12	Shahpura	6098	92.62	486	7.38	6584
13	Viratnagar	4016	87.21	589	12.79	4605
Total		132533	83.44	26301	16.56	158834

Source : District **Collector** Office, Jaipur.

In Jaipur district 79.61 percent wells were not provided with any kind of cover[12]. This adds to pollution in wells with dust and leaves of the trees nearby. Plantation of tree like babul, pipal or neem near the drinking well, is the most common traditional custom, and still prevailing. Due to absence of any covering near the wells, excreta of birds and sometimes birds also fall down inside the wells and cause contamination of water. Water-washed diseases are caused by such type of improper arrangement. Wells lacking suitable cover and platform also add pollution from those persons who come for fetching water at these wells. In this way, waste water and other impurities, e.g., detergents, dyes, etc. get mixed in the well-water.

6.3 PROTECTIVE MEASURES FOR DUG-WELLS

Although it is generally possible to take special care in the construction and protection of shallow dug-wells to render the water safe for drinking purposes, yet the possible means by which pollution may be brought about are so numerous that it is usually advisable to regard such a well with suspicion until it has been properly examined and declared safe. More care should be taken in the selection of a well-site so as to reduce the possibility of contamination to the minimum. The surface of the ground immediately around the well should be drained properly; furthermore, the vicinity of the well should be carefully protected from general surface drainage (Table 6.4).

Table 6.4 : Recommended safe-distance between wells and sources of pollutions[13]

S. No.	Sources of Pollution	Distance in metres
1	Pipes with water tight joint	3
2	Other pipes	15
3	Septic tanks	15
4	Percolation Zone	30
5	Sewage farms	30
6	Infiltration ditches	30
7	Dry wells	15
8	Cesspits	45

A well on low ground near a river bank should be protected from the risk of damage; the coping should be water-tight and should stand 2-3 m. above the ground level. The well should be maintained as far as possible from cesspools, surface drains, refuse damps and other drains which offer possibility of contamination from polluting matter.

The character and structure of the underlying rocks are the basic factors in the safety of a well; it is easily possible for a well to be polluted from a source lying on lower ground. An apparently safe well may become contaminated by the reversal of the underground drainage owing to drought or heavy pumping, and it is only when the water in the well is always higher than any possible source of pollution, that the well is safe. In flat or gently sloping ground heavy pumping seriously depress the water tables and thereby alter the direction of flow of the underground water. This may cause polluting matter to reach the well from places that in ordinary circumstances would be at a safe distance. Where the upper layers are clayed and impervious, polluting matter thrown on the surface could not reach the water-table, rather it will be carried away by the first rains. Unless traversed by cracks or other openings, a thickness of about 5-10 m of soil, loam or fine sand would be removed by natural filtration, all or most of the dangerous organic matter from descending water. But where the beds are of coarse sandy character and highly porous, polluting material would reach the water table almost unaffected by its descent, and followed by transmission directly to the well water.

The distance between a well and sources of pollution is one of the responsible factors determining the amount and nature of the pollution. Moreover, the pollutants entering from the surface or gaining access from a cesspool are more dangerous. In reference with safe-distance no absolute radius can be laid down for wells (Table 6.4), rather for all case individual attention is required. The safe-distance in any particular instance depends, for example, upon the geological character of the rocks, the depth to water level and the slope of the ground. But in fine-grained rocks like sand, sandstone, sandy clay, shale, etc. free from open joints or other passages, about 45 m is generally a safe-distance[14], if water has to percolate through the body of the rock. Dixey[15] records a case in which a well in sandy soil was contaminated by sewage at a distance of 450 m. A well in which water rises suddenly after heavy rains can have no other interpretation than that the water reaches the well without being filtered through a large depth of soil. An abrupt rise and fall of a few centimetres might have no significance, but where there is a rise and fall of a metre or more, there is no doubt that the well is liable to yield unsanitary water. Sudden appearance of much silt in the water, or of leaves, twigs and other rubbish, would also indicate that

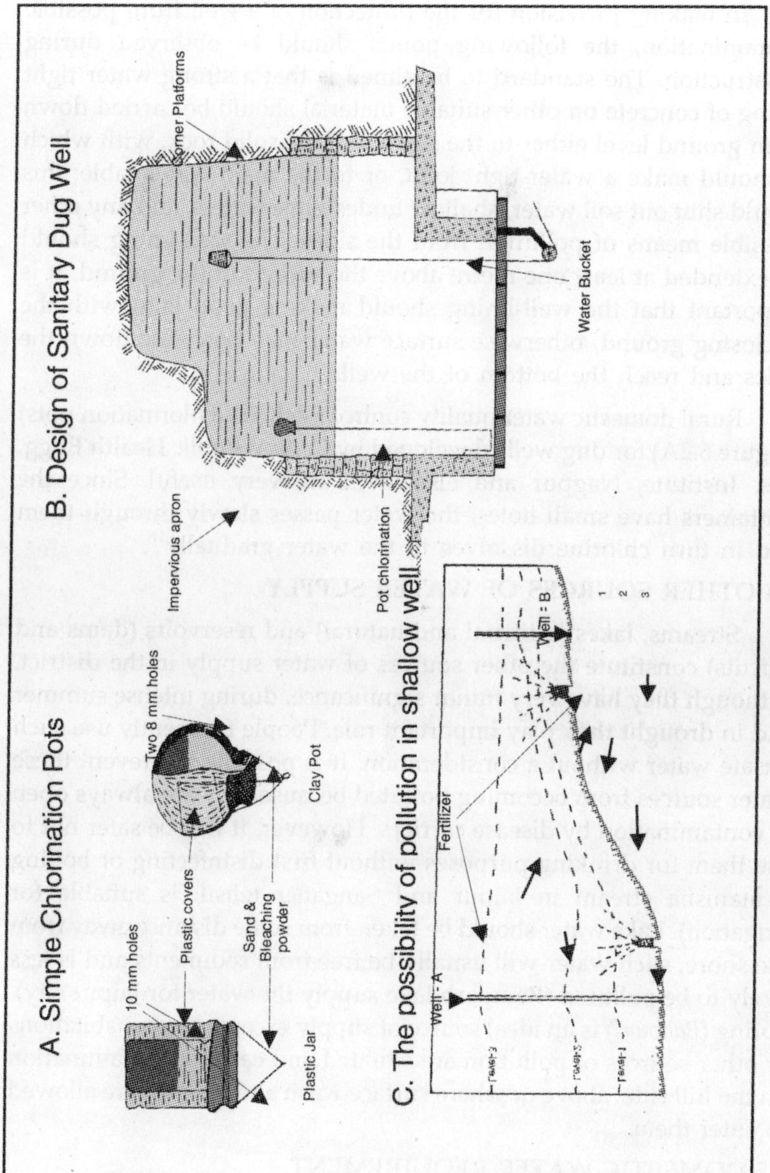
A. Simple Chlorination Pots
10 mm holes
Plastic covers
Sand & Bleaching powder
Plastic Jar
Two 8 mm holes
Clay Pot
B. Design of Sanitary Dug Well
Impervious apron
Corner Platforms
Water Bucket
Pot chlorination
C. The possibility of pollution in shallow well
Fertilizer
Well - A
Well - B

Figure : 6.2

water is reaching the well by some channel which does not provide effective filtration.

In making provision for the protection of a well from possible contamination, the following points should be observed during construction. The standard to be aimed is that a strong water tight lining of concrete on other suitable material should be carried down from ground level either to the surface of the solid rock, with which it should make a water-tight joint, or to the main water table; this would shut out soil water, shallow undesirable waters, and any other possible means of pollution from the sides. The well-lining should be extended at least one metre above the surface of the ground. It is important that the well-lining should make a good joint with the enclosing ground, otherwise surface water may percolate down the sides and reach the bottom of the well.

Rural domestic water quality control methods (chlorination pots) (Figure 6.2A) for dug wells developed by Central Public Health Engg. Res. Institute, Nagpur and elsewhere are very useful. Since the containers have small holes, the water passes slowly through them and in turn chlorine dissolves in the water gradually[16].

6.4 OTHER SOURCES OF WATER SUPPLY

Streams, lakes (artificial and natural) and reservoirs (dams and anicuts) constitute the other sources of water supply in the district. Although they have very minor significance, during intense summer and in drought they play important role. People frequently use such unsafe water without a consideration. It is possible to prevent these water sources from becoming polluted because they are always open to contamination by disease carriers. However, it will be safer not to use them for drinking purposes without first disinfecting or boiling (Amanisha stream in Jaipur and Sanganer tehsil, is suitable for irrigation). Lake water should be taken from some distance away from the shore, such water will usually be free from sediments and is less likely to be polluted (Ramgarh lake supply the water for Jaipur city). Spring (*Bahwwri*) is an ideal source of supply except where, habitations or other sources of pollution are situated and causing contamination on the hill side, above or where surface wash and rubbish are allowed to enter them.

6.5 DOMESTIC WATER REQUIREMENT

There is a wide variation in consumption pattern of water for domestic uses depending upon season, believes and customs and habits of the consumers. Specially for rural area there is no norm of water consumption and on the other hand people are not aware. It has been estimated that a person in the area requires about 60 litres of water per day throughout the year with slight variation between summer and winter season. According to PHED, Jaipur (Table 6.5) per capita about 70 litres water is required in the rural areas[17].

Table 6.5 : Norms for Providing Drinking Water in Villages

S. No.	Purpose	Litres per head per day
1	Drinking	3
2	Cooking	5
3	Bathing	15
4	Washing Utensils & House	7
5	Ablution	10
6	For Cattle	30
	Total	70

Source : PHED, Jaipur.

Presently, large numbers of villages receive potable water through supply schemes provided by PHED at the rate of 40 litres per head per day and some villages of Dudu and Phagi tehsil receive less than 35 litres per head per day[18]. But this amount of water is not sufficient as reported by villagers and even they have to depend upon the dug-wells for the rest of their water requirements. It is notable that in those villages where PHED's water supply is available, the dug-wells are seldom well maintained, although such wells still provide water in the crisis period. If people have very little water either because there is extremely little water available or because it is so far away in a well that the effort to bring it home is very difficult, then it may not be possible to maintain reasonable personal hygiene. There may be too little water for washing oneself, food, utensils or even clothes. Remaining unwashed not only allow skin infections to develop but also makes it easier for intestinal infections to spread from one person to another due to dirt. Practically, these are an important group of diseases which are very common and to be called the water-washed infections as they result from lack of water for washing or personal hygiene. Clearly their prevention depends on availability,) and access to adequate quantity of domestic water.

6.6 PROBLEMS ENCOUNTERED IN WATER SUPPLY IN HIGH CHEMICAL QUALITY AREA

In case of rural areas wherever quality parameters of ground water permits construction of handpumps, it is suggested that the coverage should be done by providing handpumps. It is only in respect of such villages which have quality problem (details in previous chapter) that it is suggested that these should be covered by regional schemes. The regional schemes for these area will in fact be the extension schemes of the main village. Since the fluoride problem has been given highest priority in case of Jaipur district the regional water supply schemes for solving the problem should be executed on top priority.

REFERENCES

1. Ghosh, G. et al. (1995) : *Water Supply in Rural India,* Ashish Publishing House, New Delhi, p. 9.
2. Goument, C.E.V. (1920) : "Water Supply", *Sanitary Engineering,* Pt.1 Govt. Press (U.P.)
3. Chandran, R. (1977) : "The World Running Out of Water", *The Illustrated Weekly,* Vol.24, p.15.
4. Ghosh, G. et al. (1995) : *Water Supply in Rural India,* Ashish Publishing House, New Delhi, p. 10.
5. PHED (1988) : *Status of Drinking Water Supply in Rajasthan,* Water Mission Report, PHED, Jaipur, Rajasthan, p. 4.
6. PHED (1988) : *Ibid,* p.8
7. PHED (1997) : *An overview - Water Supply & Sanitation,* PHED, Jaipur, p.1-2.
8. Website : *www.wsp.org/pdfs,* Jalvaani, News letters on Rural Water Sanitation in India.
9. PHED (1988) : *Status of Drinking Water Supply in Rajasthan,* Water Mission Report, PHED, Jaipur, Rajasthan, p. 27.
10. Ghosh, G. et al. (1995) : *Water Supply in Rural India,* Ashish Publishing House, New Delhi, p. 176.
11. Ghosh, G. *et al.* (1995) : *Ibid,* p.164.
12. Website : *www.wsp.org/pdfs,* Jalvaani, News letters on Rural Water Sanitation in India.

13. Romero, J.C. (1970) : *Ground Water,* New Orleans, Vol. 8, No.2, p. 37-48.

14. Fuller, M.L. (1910) : *Underground Wastes for Farm,* U.S. Geol. Survey, W.S. paper, p. 255.

15. Dixey, F (1950) : *A Practical Handbook of Water Supply,* Thomas Murby & Co., London, p. 304.

16. CPHERI (1972) : *Disinfections of Well Water (Technical Digest no. 26),* Nagpur, Central Public Health Engg. Res. Institute, Nagpur.

17. PHED (1998) : *An overview - Drinking Water,* PHED, Jaipur, p.54

18. PHED (1998) : *Ibid,* p 57.

□□□

7

Water Supply for Agriculture

Rajasthan is predominantly an agrarian State and over 70% population is connected with the agriculture sector.[1] Water is essential for agriculture in arid and semi-arid climatic condition for the proper growth of vegetables and crops. Plants consume water from the soil by absorption through their roots whose vertical length varies greatly with the species. Agriculture is the science of cultivating soil in order to produce economic crops. It demands great knowledge and skill in its scientific, commercial and artistic aspects. Very broadly, the term "agriculture" is used to include both pastoral and arable farming even though the trend is towards the latter.[2]

The relation between crop-growth and water consumption is inseparable in any agricultural planning; water needs to be supplied several times at different intervals in huge quantities throughout the crop growth. The water requirement of various crops grown in the area not only varies from season to season but different crops in the same season require varying amounts of additional water even in a year of normal rainfall. Without assessing the water requirement for the agricultural purposes, it will be useless to plan for other related aspects.

7.1 LAND UTILISATION

The land use of a region is the result of the interactions of various factors such as physical, economic and social and shows the extent to which man has been able to utilise the land resources gainfully.[3]

Table 7.1 : Land Utilisation in Jaipur District. (1995-2000)

(Area in Hectares)

Years	Total Area	Forest	Not available for cultivation			Other uncultivated land (Land Excluding Fallow Lands)				Fallow Lands			
			Area under non-agricultural uses	Barren and unculturable land	Total (4+5)	Permanent Pasture and other grazing land	Land miscellaneous tree crop & groves	Culturable Waste	Total (7-9) Land	Fallow Lands other than	Current Fallow Current Fallow	Total (11-12)	Net area sown
1	2	3	4	5	6	7	8	9	10	11	12	13	14
1995-96	1106144	71746	78241	54082	132323	78849	1273	33039	113161	58905	73704	132609	656305
Area (%)	100	6.49	7.07	4.89	11.96	7.13	0.12	2.99	10.23	5.33	6.66	11.99	59.33
1996-97	1106144	72074	78447	54709	133156	78674	1291	32376	112341	54405	48770	103175	685398
Area (%)	100	6.52	7.09	4.95	12.04	7.11	0.12	2.93	10.16	4.92	4.41	9.33	61.96
1997-98	1106083	73747	78415	53941	132356	68562	1071	31953	101586	44789	43358	88147	699247
Area (%)	100	6.67	7.09	4.88	11.97	6.20	0.10	2.89	9.18	4.05	3.92	7.97	63.22
1998-99	1105554	76851	77609	52645	130254	78345	1009	29357	108711	41874	51882	93756	695982
Area (%)	100	6.95	7.02	4.76	11.78	7.09	0.09	2.66	9.83	3.79	4.69	8.48	62.95
1999-2000	1105559	79333	81123	48326	129449	78428	953	30103	109484	51361	69627	120988	666305
Area (%)	100	7.18	7.34	4.37	11.71	7.09	0.09	2.72	9.90	4.65	6.30	10.94	60.27

Source : District Collector Office, Jaipur.

The pattern and system of agricultural land use are directly linked with the water availability in reference with its optimal quantity and required period; for example, in areas endowed with adequate water resources, lands like barren and fallows might easily be transformed into cultivated land. Land utilisation is the satisfaction from which the farm population derived the type of agriculture and developed the provision for future production and contribution to National needs[4]. Furthermore, the amount of rainfall and its distribution has a direct bearing on the growth and maintenance of vegetation. With this consideration, general land use has been discussed followed by cropping pattern and the applied aspects of water supply in this context. Some issues of water has already been referred in the context of general land use.

Land use has been classified into the categories by the division as forest, not available for cultivation, other uncultivated land, fallow land and net area sown (Table 7.1).

7.1.1 Forest Area

Our ancestors, had rightly understood the importance of forest in maintaining the environmental equilibrium. Such attitude towards trees, can certainly ensure the durability of our species[5]. The total area under forest in Jaipur district was 7.58 percent in year 2000 (Table 7.1). It has slightly increased in last five years. The mitigative measures by forest department should be undertaken in the catchment area of river basin. Because loss of forest area causes a number of difficulties, like frequent famine, less rainfall, less recharge of ground water aquifer, soil erosion, increased temperatures, pollution and many more human problems like fuel, timber, fodder, underground water shortages etc., to solve all these problems wide scale forestry work must be undertaken.

The maximum 37.19 percent forest area was found in Viratnagar tehsil followed by 23.40 percent in Jamwa tehsil. Both tehsil have maximum area under hills and all hills are under forest cover. The minimum area was found in the fluoride and salinity affected part of the Jaipur district i.e. Chaksu (0.11 percent), Phulera (0.12 percent), Dudu (0.56 percent) and Phagi tehsil (0.59 percent).

7.1.2 Not Available for Cultivation

Not available for cultivation area has been classified into the sub-division as non-agricultural uses area and barren/ un-culturable land. This area in the district was 11.71 percent. It is found that the blocks situated along the rivers basin, recorded higher percentage of non-cultivable land like in Jamwa Ramgarh (16.86 percent), Kotputli (13.41 percent) and Bassi tehsil (13.02 percent). Maximum non-agricultural area is found in Jaipur tehsil (29.52 percent) because of the effect of the capitalisation.

7.1.3 Other Un-cultivated Land

This category has been classified into sub-division as "permanent pasture/ other grazing land", "miscellaneous tree crop/ groves" and "culturable waste land". Wasteland is that land which is presently lying unused or which is not being used to its optimum potential due to some constraints.[6] The percentage of other uncultivated land has been 9.90 percent of the total land in year 2000, at block level the maximum is recorded in southern part of the district i.e. Dudu (13.93 percent) and Phagi (13.91 percent). This area consists of accumulation of sand desert and river beds, as dund,[7] and excess of soluble salts, which contain more exchangeable sodium are called alkali soil.[8] The minimum 5.27 percent other uncultivated land was found in Kotputli tehsil followed by 6.36 percent in Chomu tehsil.

7.1.4 Fallow Land

Fallow land has been classified into sub-division as "fallow land other than current fallow" and "current fallow". Fallow land area is decreasing from year 1996 due to less amount of rainfall and less soil moisture of Jaipur district. The maximum 16.81 percent fallow land area was found in Sanganer and minimum 5.43 percent in Viratnagar tehsil.

7.1.5 Net Sown Area

Net area sown increase from 1996 (59.33 percent) to 1998 (63.22 percent) and from year 1999 (62.95) onwards start decreasing and reached to 60.27 in year 2000, it was the maximum 74.52 percent in Chomu tehsil followed by 73.85 percent in Phulera tehsil. This area is being converted wasteland into NSA (net sown area) due to more population pressure on cultivated land. Minimum NSA in the northern part of the district viz. Viratnagar tehsil (39.59 percent) and Jamwa Ramgarh (41.93) have more area under hills and undulated land. (Appendix-5).

7.2 MAJOR CROPS

In Jaipur district about 60.27 percent area is under NSA. Crops has furnished directly or indirectly a significant share in the income of the district.[9] The major crops are *Bajra, Jowar,* Wheat, Maize, Barley and Gram.

7.2.1 Bajra (Pennisetum Typhoides)

Bajra is known by various names e.g. pearl millet, spiked millet and bulrush millet. It is the most important crop of the Jaipur district and occupies about one fourth area (25.92 percent) of the total harvested area. *Bajra* is a first ranking crop in seven tehsils like Amber (35.25 percent), Chomu (30.79), Phulera (27.03), Dudu (11.10), Kotputli (35.79), Viratnagar (36.71) and Shahpura (41.52). (*Appendix-6*)

In 2000 high intensity (above 30 percent) of *Bajra* was concentrated in two patches viz. middle part and northern part of the district. The middle part has sandy soil and low rainfall. Therefore, *Bajra* is the most suitable crop for this area. In north part of district, *Bajra* is common crop on undulated foot hills. Area of medium intensity (20 to 30 percent) is concentrated in two patches viz. western part and eastern part of the district where "Chomu aeolian soil" and "Bassi Rajori soil association" are found. Very low intensity (less than 20 per cent) of *Bajra* is found in the southern belt of the district due to clay soils.

Bajra has increased from 25.63 percent to 28.76 from during 1995-96 to 1996-1997 and decreased from 28.76 percent to 25.92 percent during 1996-97 to 1999-2000 (Table 7.2).

7.2.2 Wheat (Triticum aestivum)

Wheat is the first ranking crop of *rabi* season in the district. It is grown on various soils and is a very delicate and sensitive crop. It constitutes an important part of the district's trade and commerce. The crop is irrigated to get more yield and production. Wheat is produced on 24.04 per cent of the total cropped area. The maximum concentration of wheat is in the central and eastern parts of the district due to peneplain, fertile Bassi Rajori soil association, high to very high irrigation intensity and also more ground water potentiality. Medium intensity (20 to 30 percent) of wheat is found in all most north-western part, north part and north-eastern of the district. In this part, the area under wheat is medium due to hilly tract. Low and very low areas under wheat are found in the south-western parts of the district, where annual average rainfall is the lowest in the district. In this area soils are "Naraina Dudu Soil association" and "Rajori Chaksu Soil association". In this part of the district irrigation intensity is also low to medium except Phulera tehsil due to brackish ground water.

Table 7.2 : Year-wise Area Under Major Crops in Jaipur District (1995-2000)

Area in Hectares (Percentage)

	Bajra	Jowar	Wheat	Maize	Barley	Gram
1995-96	191637	15879	160340	6372	45979	61767
	(25.63)	(2.12)	(21.45)	(0.85)	(6.15)	(8.26)
1996-97	237314	21624	171382	8924	42682	66847
	(28.76)	(2.62)	(20.77)	(1.08)	(5.17)	(8.10)
1997-98	219080	16605	192549	11065	56549	142918
	(24.57)	(1.86)	(21.59)	(1.24)	(6.34)	(16.03)
1998-99	204150	17174	195548	10293	52665	155694
	(23.46)	(1.97)	(22.47)	(1.18)	(6.05)	(17.89)
1999-2000	203233	16469	188490	7706	51313	76647
	(25.92)	(2.10)	(24.04)	(0.98)	(6.55)	(9.78)

Source : *District Collector Office, Jaipur.*

During the period 1996-97 to 1999-2000 the area under wheat increased from 20.77 percent to 24.04 percent (Table 7.2) due to maximum electrification of wells, more use of high yield variety seeds and fertilizers.

7.2.3 Gram (Cicer arietinum)

Gram is the third important crop in the district. It occupies 9.78 percent area of the total cropped area in 1999-2000. It has great importance in the cropping pattern, since it increases the nitrogen content in soil and provides protein to the rural population. It needs less moisture, less care and few ploughings then wheat and give good yields on light soils.

In 1999-2000, its area of high concentration (above 12 percent) was in the north-eastern part and western part of the district. It is grown on loamy sand to sandy loam soils of "Bassi Rajori Soil Association" and Chomu Aeolian Soil Association". In this district, it is generally produced on unirrigated soils and its yield depend on winter rains. The area of low concentration are in Amber and Kotputli tehsil. Very low concentration are found in the middle and south-eastern parts of the district.

In 1995-96 gram occupied 8.26 per cent of the harvested area. It was an important crop in the northern and eastern part of the district. The area under gram has increased from 8.10 percent to 17.89 percent during the 1996-97 to 1998-1999 in the district. In year 1999-2000 its area has decreased 9.78 percent and the area under gram has decreased due to the improvement in irrigational facilities and high yield variety (HYV) seeds of wheat. Now wheat is replacing gram in the tehsils where irrigational facilities are increasing.

7.2.4 Barley (Hordeum vulgare)

Barley is the fourth important crop in the district. It occupies 6.55 percent to the total occupied area. Barley matures in a short period and requires less irrigation than wheat. It has shallow roots, and is generally produced on light soils which are not suitable for wheat cultivation. It is also used as food and fodder crop.

Maximum concentration of barley is in the western part of the district, where it occupies more than 10 percent of the total cropped area. Soils of this part belongs "Chomu Aeolian Soil Association" and these soils are generally sandy loam to clay loam with deep calcareous sub soil. These soils are suitable for barley crop.

Medium intensity (4 to 8 percent) of barely is found in all most north-eastern and middle of the district. Soils of this part belongs to loam to clay loam of "Bassi Rajori Soil Association" category, but generally it is grown on poor soils. Low and very low areas under wheat are found in the south-western parts of the district, where annual average rainfall is the lowest in the district. Barley slightly increased from 5.17 percent to 6.55 percent during 1996-97 to 1999-2000.

7.2.5 Jowar (Sorghum vulgare)

The word sorghum is derived from the latin word "sorgo"

meaning "rising above", that is, above other crops in the arable fields by its side. It is also referred as the "giant or great millet" being the largest of the food- grains called millets both in terms of plant and grain.[10] It is drought resistant crop, has the ability to become practically dormant during periods of severe droughts and growth, without much apparent injury until the conditions become more favourable.[11] The main concentration of *Jowar* is in the areas of heavier soils. It is produced in rotation with gram in some areas where *Jowar* fields are kept fallow in next *rabi* season.

On an average there was 2.10 percent of the total cropped area under *Jowar* in 1999-2000. The maximum area (more than 7 per cent) under *Jowar* is concentrated in the south-western part of the district in Dudu (8.57 percent) and Phagi tehsil (8.00 percent) where soils are sandy loam to loam surface texture of "Naraina Dudu Soil association" and "Padasoli Phagi Soil association". The south-western part has saline and alkali soils. *Jowar* tolerates alkalinity or salinity better than other crops. Low and very low areas under *Jowar* are found in the north-western part, middle part and north-eastern of the district, where it was replaced by wheat.

Area under Jowar is almost same from 1999-96, it is slightly increased from 2.12 percent to 2.62 percent during 1995-96 to 19996-97, slightly decreased in 1997-98 and again increased from 1.86 percent to 2.10 percent during 1997-98 to 1999-2000.

7.2.6 Maize (Zea mays)

Maize is the second most important *kharif* crop and is a popular winter staple food of inhabitants. It makes specific moisture demands during its period of growth, and the crop is very conservative in the use of water during its early phases of developmental.[12] Excessive rainfall and drought, both damage the standing crop. It is an exhausting crop, so even fertile soils also require manure or chemical fertilizers for per hectare high yield. Its cultivation requires more attention and careful culture in the form of manuring, weeding and thinning. It is a form of intensive method of cultivation.

Maize occupied only 0.98 percent of the total harvested area in 1999-2000. Maximum concentration of maize is in the Jamwa Ramgarh tehsil (4.13 percent), followed by Dudu tehsil (2.34 percent) (Appendix-6). Medium intensity (0.5 to 1.5 percent) of maize is found in almost south-eastern and middle part of the district. Maize is grown only in those areas where soils are sandy loam to loam surface. The area under maize also slightly increased from 0.85 to 1.18 percent during 1995-96 to 1998-99 (Table 7.2) and slightly decreased in year 1999-2000.

7.3 DOUBLE CROPPING

The availability of water holds a quite positive relationship with proportion of double-cropped area to the cultivated area. As such, the evaluation of general land use of the study region will remain incomplete unless the distributional pattern of double cropped area is considered properly. Besides irrigation, a few other factors like per capita availability of cultivated land among the rural population, existence of enthusiasm among the cultivators towards improved methods of farming, the nature of terrain and its level has also no less importance in controlling the double cropping practice. The fact, that the surface level plays an important role in the context of double-cropping may be elaborated with examples of low lying area which suit only the cultivation of single crop, i.e *Bajra* in the Kharif season. It is obvious from the Appendix -8 that the proportion of double cropped area exhibits great variations. The percentage of double cropped to the gross cropped area above twenty five percent has been recorded in four tehsils, i.e. Chomu (28.20 percent), Kotputli (29.05 percent), Viratnagar (26.66 percent) Shahpura (27.65 percent).

7.3.1 Cropping Intensity

The cropping intensity gives an idea of the extension of double cropped area on the net sown area or as a percentage of the total cropped area. Cropping intensity deals with the use of cultivated land within one agricultural calendar year.

There is very limited scope for increasing the net sown area in the district. In these circumstances farmers should produce more than one crop from the same field within one agricultural year and this will increase cropping intensity. High cropping intensity is the only solution to the problem of feeding the increasing population. The per hectare yield can be increased by adopting improved agricultural practices i.e. use of water resources, chemical fertilizers, high yielding variety of seeds, modern agricultural implements etc. The cropping intensity depends on irrigation intensity, soil characteristics i.e. soil types, water availability, rainfall etc. For determining the cropping intensity following formula is used :

$$\text{Index of cropping intensity} = \frac{\text{Total cropped area}}{\text{Net sown area}} \times 100$$

The cropping intensity is very high in one belt. It is in the northern part of the district. In this belt soils are mainly "Bassi Rajori Soil Association" and Chomu Aeolian Soil Association. Besides, this area has high ground water potentiality. In year 1999-2000 high cropping intensity was found only in Kotputli tehsil (Appendix - 8). The medium intensity is concentrated in the middle part and north-eastern part of the district. In this part irrigation intensity is medium to high and water potentiality is also high.

Low cropping intensity is found in the southern and western parts of the district, where rainfall, irrigation intensity, arithmetic density and water potentiality are also low. This area is agriculturally most backward in the whole of the district. Therefore, cropping intensity is also very low. The cropping intensity also slightly increased from 113.91 to 127.52 percent during 1995-96 to 1997-98 and slightly decreased from 127.52 to 117.66 percent during 1997-98 to 1999-2000 (Appendix -7).

7.3.2 Cropping Pattern

The main crop of the district is the Rabi crop grown in the month of October and harvested in March. Figure 7.6 shows that the Rabi crop area have more than the "Rabi crop irrigated area" in all tehsil of Jaipur district in year 2000. The crop consists of grains like wheat, barley and gram. Wheat is sown as both irrigated and un-irrigated crop. Barely replaces wheat on the irrigated lands where the soil is light, irrigation water is scanty or its mineral content is high. When the monsoon are scanty then the area cultivated by barley increases. In the case of gram generally irrigation is not required, it is also sown together with wheat and barley.

The Kharif crop also contributes to the supply of food grain to the district. It is generally sown in May-June end and harvested in September. Kharif crop area is about three times more than the "Kharif crop irrigated area" in all tehsil of Jaipur district in year 2000. The main crops that are sown during this season are *Jowar, Bajra,* Groundnut, Maize, Pulses like *urad, moong* and *moth.*

Fruits and vegetables are grown in abundance in the areas surrounding the city of Jaipur and more so in the areas of Sanganer and Amber. The crops are grown on commercial basis to fulfil the needs of the urban area. The fruits mostly include papaya, guava, mango and citrus fruits. The source of irrigation is mostly from wells and also to some extent crops are watered by the drainage water.

7.4 IRRIGATION WATER REQUIREMENT AND IRRIGATED AREA

The irrigation requirement refers to the quantity of water exclusive of precipitation required for crop production and it also includes economically unavoidable waste. Several researches during last decades were carried out to find out the minimum water requirement of different crops in order to ensure the most economical use of water compatible with maximum yield[13]. The use of high yielding seeds together with high degree of fertilisers, has been inextricably linked with assured water resources, either through reliable natural rain water or artificial irrigation[14]. In the district, irrigation is an extremely important and effective means of increasing agricultural production as the precious source of water is unevenly distributed both in respect of location and seasons and is also erratic resulting in recurrent drought and famines[15]. It is in this context that the importance of irrigation facilities has been examined in Jaipur district despite the fact that it belongs to semi-arid region. In Jaipur district irrigation sources can be categorised into Ground water and Surface water.

7.4.1 Irrigated Area through Wells & Tubewells (Ground Water Source)

Ground water is the most important source of irrigation in this district. Ground water is utilised for irrigation through masonry wells with bullocks and motes, dug wells and dug-cum-bored wells, fitted with pumps operated by diesel engines and electric motors, and tubewells. Dug-cum-bore wells, also called cavity wells are more popular in certain patches of the district, where in sandy aquifer occurs at comparatively shallow depth. The table 7.3 shows the percentage of irrigated area by different sources in year 1996 to 2000.

The Table 7.3 reveals that 99.43 percent area was irrigated by wells and tubewells in year 2000. During five years (1996 to 2000) the irrigated area under wells increased by 4.44 percent. It further shows that the importance of irrigation is increasing continuously. The main reason of increasing irrigation by wells is the erratic nature of rainfall. After the green revolution, cultivators are becoming prosperous in the Jaipur district. Now the farmer wants his own well on his farm. Thus, the number of wells has been increasing very rapidly.

Table 7.3 : Year-wise Irrigated Area by Different Sources in Jaipur District (1996-2000)

Area in Hectares (Percentage)

Year	Wells & Tubewells	Canals	Tanks	Other Sources	Net Irrigated Area
1996	331516 (94.99)	15151 (4.34)	2303 (0.66)	24 (0.01)	348994
1997	333076 (95.20)	12640 (3.61)	4083 (1.17)	64 (0.02)	349863
1998	341522 (97.51)	7797 (2.23)	921 (0.26)	6 (0.00)	350246
1999	357349 (98.34)	5071 (1.40)	768 (0.21)	202 (0.06)	363390
2000	357191 (99.43)	1976 (0.55)	65 (0.02)	0 (0.00)	359232
Change over 1996	**+(4.44)**	**-(3.79)**	**-(0.64)**	**-(0.01)**	

Source : *Dept. of Irrigation, Jaipur.*

In the Jaipur district irrigated area by canals (0.55 percent) and tanks (0.02 percent) was very low. During last five years (1996 to 2000) the irrigated area under canals decreased by 3.79 percent and area under tanks decreased by 0.64 percent (Table 7.3). It shows the impact of drought since 2000.

Tehsil-wise irrigation by wells varies from 94.26 to 100 percent. About 12 tehsils have more than 99 percent irrigation by wells. There is only one tehsil (Sanganer), in the district where irrigation by wells is less than 95 percent (94.26 percent) of total irrigated area (Appendix- 9).

7.4.2 Scope of Irrigation for Ground Water Development

The total quantity of water supplied for irrigation by groundwater has been considered as ground water draft for irrigation. In Jaipur district, block-wise assessment of ground water estimation of draft has been made by Central Ground Water Board (CGWB), based on the hydraulic characteristics and on the basis of the quality of ground water, different potential zones have been delineated (Appendix-10). Ground water potential areas for irrigation are identified in the following blocks :

7.4.2.1 *Amber Block*

On the basis of hydraulic characteristics two potential zones have been identified in this block. The first zone is identified in north and western portion of the block and comprises an area of 746.46 sq.km. The main water bearing formation of the zone is alluvium and depth to water table ranges from 9.0 m. to 31 m. The average discharge of tubewells in this zone is 9000 lph.

The second zone is identified in eastern portion with an area of 104.02 sq.km. The main water bearing formation in this zone is quartzite. Depth to water table ranges from 12.7 m. to 31.40 m and the average yield of well fitted with pumpsets is 40000 lpd (Appendix-10).

7.4.2.2 *Viratnagar Block*

On the basis of hydraulic characteristics, two potential zones have been delineated in the block. The first zone intends all over the block except southern portion with an area of 588.37 sq.km. The main water bearing formation is alluvium of quarternary system. The depth of water table ranges from 3.6 m. to 23.4 m. The yield of wells operated by pumping sets is 93400 lpd.

The second zone is identified in eastern portion with an area of 75.59 sq.km. The main water bearing formation in the zone is quartzite. The depth of water below ground level fluctuates between 11.0 m. and 13.0 m. The average discharge of the tubewell in this zone is 6000 lph (Appendix-10).

7.4.2.3 *Bassi Block*

In this block, on the basis of hydraulic characteristics, three potential zones have been identified. The first zone is identified in the western portion of the block with water table depth ranging from

12.0 m. to 13.0 m. The yield of wells operated with pump sets is 65000 lpd. The second zone occupies partly eastern, southern and northern portions of the block. The depth of water table observed in this zone is 14 m. to 27 m. The average discharge of tubewells in this zone is 10000 lph.

In the central portion of the block, third potential zone is identified and depth to water table in this zone ranges from 23.2 m. to 27.9 m below ground level. The average yield of wells operated by pumpsets is estimated as 52500 lpd (Appendix-10).

7.4.2.4 *Chaksu Block*

On the basis of hydrogeological characteristics of the block two potential zones have been identified in the area. The first zone covers an area of 552.06 sq.km. and occupies a major area of the block. The principal aquifer of the zone is semi-consolidated older alluvium, compared essentially of lime. Kankar and clay with sand and silt at places. The depth of water below ground level ranges from 7.8 m. to 22.8 m. The average discharge of tubewells in this zone is 15000 lph.

The second zone occupies an area of 177.29 sq.km. in north central and south central boundary of the block. Ground water movement takes place within the weathered and permeable capping of this zone along schistosity, joints and fractured planes. The depth to water table during premonsoon varies from 13.15 m. to 17.5 m. The average yield of wells operated by pumpsets is 52500 lpd (Appendix-10).

7.4.2.5 *Dudu Block*

On the basis of hydraulic characteristics three potential zones have been identified. The areas of these three potential zones are 327.26 sq.km., 407.92 sq.km. and 1055.67 sq.km. respectively. The main water bearing stratum in these three zones are older alluvium, mic-schist and gneisses. The depth of water table ranges from 10.7 m. to 21.1 m in the first zone from 8.90 m. to 28.9 m. in the second zone and from 6.4 in to 19.6 m. in the third zone as recorded during pre-monsoon season. The average yield of wells fitted with pumpsets estimated in these zones are 57500 lpd, 38000 lpd and 38000 lpd respectively (Appendix-10).

7.4.2.6 *Govindgarh Block*

In this block, one ground water potential zone is delineated with an area of 638.08 sq.km. The zone is spread all over the block. The main water bearing stratum is quaternary alluvium mainly compared of sand, silt and clay. The depth of water table observed during pre-monsoon season varies from 9.9 m. to 31.0 m. The average discharge of tubewells in this zone is 15000 lph (Appendix-10).

7.4.2.7 *Jamwa Ramgarh Block*

Based on the hydraulic characteristics, the entire area has been divided into five potential zones. These five zones occupy the areas of 577.82 sq.km., 96.85 sq.km., 78.95 sq.km., 136.00 sq.km. and 72.25 sq.km. respectively. The main water bearing stratum of these zones are alluvium, quartzite, fractured and weathered rock, schist and granite respectively. The depth of water table ranges from 5.6 m. 28.5 m in the first zone, 13.0 - 14.0 m. in the second zone, 16.0 - 17.0 m. in the third zone, 20.5 - 25.0 m. in the fourth zone and 13.0 - 14.0 m. in the last zone. The average yield of wells fitted with pumpsets in these zones is 85000 lpd, 45000 lpd, 45000 lpd, 40000 lpd and 40000 lpd respectively (Appendix-10).

7.4.2.8 *Jaipur Block*

In this block only one potential zone have been identified based on hydraulic characteristics of the area. The zone occupies major portion of the block and the main water bearing stratum is being alluvium compared of varying proportions of silt, clay and kankar. The depth of water table below ground level varies from 10.55 m. to 18.90 m. The average yield of wells operated by pumpsets is 96500 lpd (Appendix-10).

7.4.2.9 *Kotputli Block*

In this block two potential zones have been delineated with an area of 427.32 sq.km. and 130.54 sq.km. respectively. The first zone occupies major portion of the block. The main water bearing strata in these blocks is alluvium and quartzite.

The depth of water table observed is from 6.1 m. to 14.20 m in the first zone and 13.1 m. to 27.75 m. in the second zone. The average discharge of tubewells in the first zone is 25000 lph. The average yield of wells fitted with pumpsets in the second zone is 40000 lpd (Appendix-10).

7.4.2.10 *Phagi Block*

On the basis of hydraulic characteristics three potential zones have been identified with an area of 396.11 sq.km. and 880.44 sq.km. respectively. The main water bearing stratum in the block is alluvium consisting of unconsolidated and semi-consolidated sand with clay. The depth of water table ranges from 4.5 m. - 30.0 m. in the first zone and in the second zone it varies from 15.0 m. to 16.0 m. The average yield of wells fitted with pumpsets is 60000 lpd and 43400 lpd in the first and the second zones respectively (Appendix-10).

7.4.2.11 *Sambhar Block*

The two potential zones have been demarcated in the block with areas 249.44 sq.km. and 635.55 sq.km. respectively in northern and southern portion of the block. The principal aquifer of the zone is younger alluvium and sand. The depth of water table in the first zone ranges from 11.50 m. - 13.50 m. while the same in the second zone ranges from 6.7 m. - 21.9 m. The average discharge of tubewells is 25000 lph (Appendix-10).

7.4.2.12 *Sanganer Block*

Only one potential zone is identified in this block with an area of 613.89 sq. km. which extends all over the block. The water bearing formation in this zone comprises of Alluvium with silt, clay and kankar. The depth to water table in the zone varies from 16.50 m. to 23.90 m. The average discharge of tubewells in this zone is 10,000 lph (Appendix-10).

7.4.2.13 *Shahpura Block*

In this zone also only one potential aquifer is identified which extends north to south and east to west of the block covering an area of 384.13 sq.km. The main water bearing stratum is alluvium with water table depth of 17.50 m. to 29.00 m. The average discharge of tubewells in this zone is 18000 lph (Appendix-10).

7.4.3 Feasibility of Tubewell Construction

Tubewells are feasible in the area covered by appreciable thickness of alluvium. The area of Jaipur district feasible for high capacity tubewells (more than 1,00,000 litres per hour) are restricted. and the area feasible for medium to high capacity tubewells are restricted to Chaksu and Amber block. In the remaining part of the district low to medium capacity tubewells are feasible in alluvium. In hard rock areas bore wells are feasible, where structural weak planes such as joints and fissures traversing the rock persist down to considerable depth. In the fractured hard rock, the openings are interconnected, hence a drilled well will give copious supply of water. The following areas are demarcated for the construction of tubewells of different capacities by Central Ground Water Board in Jaipur district.

7.4.3.1 *Medium to High Capacity Tubewells*

Medium to high capacity tubewells (50,000 to 1,00,000 lph) are feasible in the areas:

- around Kotkavda in Chaksu block
- around Sanganer and Dholiawas (Sanganer block)
- around Achrol in Amber block.

7.4.3.2 *Medium Capacity Tubewells*

Low to medium capacity tubewells (25,000 to 50,000 lph) are feasible in the areas:

- Around Mahana Madrampur, Nansingpura, Katipura and Goner villages (Sanganer block),
- Around Nimora, Kasliipura, Ruphari, Bidakhan and Basera villages (Chaksu block),
- Around Paliawas, Nlaharajpura, Deogaon, Bassi, Tunga, and Lowan villages (Bassi block).

In the areas where groundwater occurs in the weathered and semi-weathered mantle and joints in hard rocks, groundwater development can be taken up only by means of large diameter dugwells and also by revitalising and deepening of existing ones by boring and blasting and through horizontal boring. The area covered by hard rock aquifers possesses large number of tanks and seepage from them could be utilised as a source of artificial recharge and the development of groundwater is taken up by sinking additional wells on the downstream side of the tanks. This will help to solve water supply problem wherever faced and may be taken as remedial measure to prevent water logging conditions and formation of saline residues due to seepage of water from tanks.

In the alluvial areas the yield of dugwells can be increased by revitalising them by drilling in the bottom and installing slotted pipes against the sand zones. Further, large scale development can be done in areas adjacent to the river courses and also in the proximity of dunes, where groundwater exploitation is comparatively less.

Table 7.4 : Spacing of Tubewells

No.	Site	Transmissivity	Sp. Yield	Discharge considered	Duration of Pumping considered $m^3/d/m$	Spacing in m. 5% interference	Spacing in m. 1% interference	Remark
1	Amber	169.00	16.09	60	0.66	156	220	Site 1 to 4 falls in
2	Johtwara	179.00	4.72	60	0.66	120	220	area feasible for
3	Bassi	145.23	1.60	60	0.66	190	300	medium to high
4	Sanganer	99.88	12.79	60	0.66	154	200	capacity tubewells
5	Mohana	112.20	14.10	30	0.66	42	90	Site 5 to 6 falls in
6	Ramgarh	57.00	9.72	30	0.66	70	110	area feasible for low to medium capacity tubewells

Source : *Central Ground Water Board, Jaipur.*

7.4.4 Spacing Between Two Tubewells

This refers to the safe distance between two pumping wells to minimise the interference of one on the other. The spacing depends on the distance drawdown relation around a pumping well which in turn depends on hydrogeological parameters such as transmissivity, specific yield, delay index, and discharge of the pumping well and duration of pumping. Considering the aquifer parameters of the exploratory wells in the area, spacing has been worked out for two different sectors. This includes areas feasible for medium to high and low to medium capacity tubewells, as given in Table 7.4.

The distance-drawdown relation was established for a pumping period of 0.66 day (16 hours) using modified Prickett formula[16] :

$$\boxed{S = \frac{Q}{4 \times KD} W \times (Uv\, r/B)}$$

Where:

S = drawdown in m.

Q = discharge in m^3/d.

KD = Transmissivity in $m^3/d/m$

(Uv r/B) = Well function for water table aquifer with fully penetrating wells.

Based on above analysis, the spacing between two medium to high capacity tubewells for the discharge mentioned in table 7.4, ranges from 120 m to 190 m (with 5 percent interference). This spacing holds good for random development, as in the case of sinking tubewells for drinking water purposes installed by PHED and villagers. However, for systematic development of the area by tubewells for irrigation purposes the spacing should be more than 200 m to 300 m depending on the hydrogeological conditions prevailing in the area.

The spacing between too low to medium capacity tubewells for a discharge of 30 m^3/hr. ranges from 42 m to 70 m (with 5 per cent interference) and for systematic development of the area by tubewells for irrigation purposes the spacing should be more than 90 to 110 m.

7.4.2 Irrigated Area through Tanks/Canals (Surface Water Source)

Jaipur district is very poor in irrigated area through tanks (only

0.02 percent) and canals (0.55 percent) in year 1999 - 2000 due to drought condition. Tehsil wise irrigated area by tanks is very small in Phagi (60 hectares) and Jamwa Ramgarh (only 5 hectares) (Appendix -9).

Maximum irrigated area by canals in 1999 - 2000 was in Sanganer tehsil (5.74 percent), followed by Kotputli (0.81 percent) and Amber (0.55 percent) tehsil (Appendix-9). Irrigation by tanks and canals is very small in every tehsil of Jaipur district but it has more importance in south-western part particularly in Dudu and Phagi tehsils. Irrigated area by tanks and canals decreased from 2303 hectares to 65 hectares and 15151 hectares to 1976 hectares (Table 7.3) in 1999-2000 due to drought conditions. Tehsil wise analysis also shows that tank and canal irrigation has decreased in every tehsil.

The word tank is generally applied to artificial ponds and lakes made either by excavating or by banking. Some tanks are also made by damming the nullahs. There are 120 tanks all over the district. The largest number of tanks is found in Phagi tehsil (55 tanks), followed by, Dudu (34), Chaksu (13), Jamwa Ramgarh (6), Phulera (4). Phagi, Dudu and Chaksu tehsils lie in the southern part of the district where soils are clay and ground water is brackish in scattered patches. While Jamwa Ramgarh tehsil lies in the hilly tract where tanks can be seen in the villages located in the foot hills. The above five tehsils, which constitute one-third area of Jaipur district have 93.33 percent tanks. The remaining tehsils have very few tanks, Amber and Bassi have 3 tanks each and Viratnagar have only 2 tanks. In the five tehsils (Jaipur, Sanganer, Chomu, Kotputli and Shahpura) of Jaipur district have no any useful tank for irrigation purpose.

In the last three year few (7 tanks in 1998, 1 tank in 1999 and 4 tanks in 2000) irrigation tanks overflow in Jaipur district due to drought conditions. At present, ground water plays an important role in the development of water resources in Jaipur district[17] and the total quantity of water supplied for irrigation by ground water has been considered.

REFERENCES

1. Department of Irrigation (1996): *Report of the Working Group on Irrigation for The Ninth Five Year Plan (1997-2002)*. Department of Irrigation, Govt. of Rajasthan, Jaipur.

2. Somani, L & Tikka, S.S. (1983) : *Dictionary of Agriculture*, Agricole Publishing Agency, New Delhi. p. 292.

3. *Donald, A.D. (1982) : Soil and Land Use Planning,* London. p. 49.

4. Buck, J.L. (1937) : *Land Utilisation in China, Vol. 3.* London. p. 63.

5. Ghosh Shanker (1976) : "The Planning of our Forest Resources", *Yojna,* Vol. XX/16, p.10.

6. NRSA (1980-82) : *Mapping of Wastelands in India from Satellite Imagery, Summary Report,* Rajasthan, Hyderabad.

7. *Ibid* (1980-82).

8. *Ibid* (1980-82).

9. Website : *http://www.rajamb.com/soil.htm*

10. Singh, J.B. (1976) : *An Agricultural Geography of Haryana,* Vishal Publication, Kurushetra, India, p. 378.

11. Singh, J.B. (1976) : *Ibid.* p. 378.

12. Singh, J.B. (1976) : *Ibid.* p. 285.

13. Sally, H.L. (1968) : *Irrigation Planning for Intensive Cultivation,* Asia pub. House, Bombay, p. 69-94.

14. Mukeherji, B.K. et. al. (1962) : *Review of Work Done on Water Requirement of Crops in India,* Indian Council of Agriculture Research, New Delhi, p 31.

15. Dept. of Agriculture (1995) : *Report of the Committee for the Fourth Five Year Plan on Agriculture Production - Rajasthan State,* Part I. p. 10

16. Prickett (1965) : "FAO Meeting Report", Food and Agriculture Organisation of the United Nations World Health Organization, 1965/10/1, p.27

17. Dept. of Irrigation (1996) : *Report of the Working Group on Irrigation for the Ninth Five Year Plan,* Dept. of Irrigation, Jaipur.

❑❑❑

8

Impact of Droughts on Water Supply

Drought is a condition of water scarcity in which existing sources are unable to meet adequate or even minimum demand of the users in a particular region for that period of time. This means that drought is a situation in which the demand for water exceeds the supply from all sources. The demand for water also varies with time as well as level of technology used by the people and, therefore, severity of drought changes with time even if the same amount of water supply is maintained[1].

Rajasthan once again is enforced to face severe drought in 2002. The droughts have been the recurring problem of this country and will continue to be so for many years to come. Lack of adequate rainfall and skewed rainfall pattern has led to famine, and drinking water shortages, which in turn will affect the agricultural output.[2]

Drought causes financial as well as loss of human and animal life. Since the beginning of history, man has been seeking continually for better supply of water, and also for protection from uncontrolled water. The scarcity of water due to the failure of the rains in the first half of 1999 and wells going dry since the close of the year, created extreme hardship to the people of the area. Information gathered from different source of government (Agriculture, Relief, Ground Water and Irrigation department) suggests *that this year's drought is much more severe than the last two years*[3].

The status of drinking water and availability of fodder are

below satisfactory norms and as we enter the hot months of summer, the complexity of situation is going to worsen further to a great extent. Livestock has always been a vital component of State's economy and with lack of water and fodder, the rate of animal death is increasing rapidly. Migration of cattle owners in search of fodder and water for their animals is becoming a common phenomenon in drought affected areas. The situation of drought has been extremely critical in the southern and western part of the district. Its characteristic features are: low and erratic rainfall; low humidity and high evaporation demand; high solar radiation and a strong wind regime; erodible, sandy soils in the form of dunes and hummocky plains; deep seated, often saline groundwater; a sparse vegetation cover with reasonable floral diversity (nearly 500 species); but greatly transformed by human activity which dominates the natural endowments. High density of human and livestock population and a large incidence of arable farming compared to arid tracts of the world, besides a rich cultural heritage are its outstanding features. The high and yet rising biotic pressure and socio-political environment have interacted with natural endowments to accelerate the process of natural resource degradation with a variable impact on different strata of a rather diverse society.

8.1. CLIMATE AND RELATED ASPECTS

There is semi-arid type of climate all-over the district. The climatic conditions of Jaipur district are briefly discussed in previous chapter. Moreover, droughts and floods together sap the economic health of the region. What the region suffers from most is not as much. The lack of adequate amount of rainfall is not as much a causative factor as is its ill distribution, both in space and time.

Actual annual rainfall has been very low in district for last three years and has continuously decreased since 1999 (*Appendix 1*). In last three years (1999 to 2001) the actual annual rainfall is less than the normal annual rainfall. The rainfall during the period of June to September constitutes nearly 90 percent of the annual rainfall. The variation of the annual rainfall from 1999 to 2001 is very large. Difference with normal and actual rainfall of the district is very high in last three years. In the three years period (1999-2001) the actual annual rainfall is less than normal annual rainfall in all tehsils of Jaipur district excepted Viratnagar in 1999 and J. Ramgarh in 2000. In 2001 actual rainfall is very low in all the tehsils of Jaipur district.

In spite of precipitation being the ultimate source of water supply to a region, it is not the sole criterion for the assessment of the water potentialities. It is prerequisite to know how the precipitation water is matched against the water needs of an area; since it is very rare that precipitation to be the same as the water needs; where it is higher, the region is considered wet, while where lower, it is dry. But this simple relationship is often misleading, for the, relative seasonal marches of the precipitation, and the water need are never usually parallel, occurrence of crossovers being as common as it is large[4]. In some places more precipitation falls month after month than the water need, the accumulated surplus moves through and over the ground to streams and rivers. In others, precipitation may be deficient in one season and excessive in another resulting to a period of dryness, followed by one with runoff.

8.1.1 Tehsil-wise Position of Rainfall in (1999-2001)

The corresponding position of the year 1999 to 2001 have also been given for comparison, it can be said that monsoon was unfavourable to south-western part of the Jaipur district, for two consecutive years (2000 & 2001) and particularly unfavourable to all tehsils of Jaipur district as these were in the category of deficit for three consecutive year (1999, 2000 & 2001).

Percentage deviation of actual rainfall from its normal rainfall and it's distribution over the time period broadly adjudges the progress and quality of rainfall at any place. (The normal rainfall of all the district H.Q. and tehsils/ sub-tehsils have already been worked out by Indian meteoology Department)[5]. The deviation/ departure of actual rainfall against normal rainfall has been classified, as under, by adopting the criteria mentioned in the table 8.1 and a comparison with last two years has also been given[6].

Tehsilwise, difference with normal and actual rainfall of the district is high in year 1999. Minimum difference (above -5 cm.) of rainfall is in the northern and western parts of the district and maximum difference (less -15 cm.) of rainfall is in middle part of the district. This difference continually increased in year 2000 and 2001. Maximum effect of the low rainfall is in middle part, south-eastern part and south-western part of the Jaipur district.

Table 8.1 : List of Tehsils under Deficit/Scanty Rainfall Categories (2000-2001)

S.No.	Category	Criteria (Deviation from Normal)
1	Abnormal	60% more
2	Excess	Between 20% to 59%
3	Normal	Between (-) 19% to 19%
4	Deficit	Between (-) 59% to (-) 20%
5	Scanty	(-) 60% or less

S. No.	Tehsil	2000 Deviation %	2001 Deviation %
1	Jaipur	-40	-39
2	Bassi	-36	-33
3	Chaksu	-31	-40
4	Sanganer	-25	-
5	Amber	-50	-35
6	J. Ramgarh	-	-22
7	Chomu	-33	-
8	Samod	-36	-
9	Phulera	-46	-
10	Naraina	-59	-
11	Dudu	-52	-44
12	Phagi	-49	-
13	Shahpura	-30	-25
14	Viratnagar	-21	-

Source : *Irrigation Department, Govt. of Rajasthan, Jaipur.*

8.2 POSITION OF RESERVOIRS IN DROUGHT YEARS

As far as the position of water available in reservoirs is concerned, it is clear from the above facts that there was no good/heavy downpour which could have resulted in surface flow to feed the reservoir. In 2000 by the end of 31st December only 4 tanks had overflown, but in 2001 no one tank have overflown out of the 111 tanks in the Jaipur district. The percentage of water received upto 31st December, 2001 in major tanks was only 6.52 percent (Table 8.2) and overall tanks (more than 150 Mcft. capacity) of Jaipur district received only 13.14 percent of water.

Table 8.2 : Position of Water Received in Tanks 2001 of Jaipur district

No. of Tanks	Total Live Storage	Total Storage Received	Received Percentage
111	10627 Mcft.	693 Mcft.	6.52%

Source : *Irrigation Department, Govt. of Rajasthan, Jaipur.*

8.3. WATER SUPPLY IN DROUGHT CONDITION

Jaipur district is one of the scarcity pockets of the State and shortage of drinking water has repeatedly been acute during summer. Open insanitary wells, which serve as source of drinking water in rural areas of the district, have a tendency to dry up during summer either due to failure of required timely rains or due to general lowering down of water table due to uncontrolled draw-off from the numerous State tubewells and private pumping sets constructed for irrigation purposes. This problem is specially acute in southern part, middle part and northern-western part of the district.

The water level in these areas has gone down beyond 10 to 15 metres depth in every year. This change of the water level has been shown in table 8.3.

Thus, most of the dug-wells, which are lesser than this depth, dry up, which further worsen the situation during the summers, particularly if there is a failure of rain as well. The villagers move with big pots on their heads in search of water. They obtain drinking water either from the State and private tube-wells and pumping sets which are mostly situated outside the settlement or from ponds, rivers and reservoirs where accumulated raw water is still available. This situation furthermore becomes acute, if any of these tube-wells become out of order either due to power shortage or sometimes even due to mechanical fault.

Table 8.3 : Depth to Water Level (1993-94) and Change in Water Level (1994-2001)

No.	Range in (m)	Area
		Depth to Water Level (1993-94)
1.	< 10	Part of Govindgarh and Shahpura and southern peripheral region has shallow water level less than 10 m.
2.	10 to 20	Major part of the district has water level between the range, southern peripheral area and pockets scattered in different parts do not lie in the range.
3.	> 20	Area around Jhotwara, Govindgarh and east of Shahpura has deep water level ranging more than 20 m.
		Change in Water Level *(1994-2001)*
1.	0 to 3	Area around Dudu and Phagi situated in southern part of the district exhibit marginal rise in water level.
2.	0 to -3	Southern and north-western peripheral region show marginal depletion in water level between the range.
3.	-3 to -7	Major part of the district excluding southern peripheral region and pockets located in different part exhibits depletion in water level between the range.

Source : State Remote Sensing Application Centre, Govt. of Rajasthan, Jodhpur.

The shortage of rainfall during the last three years has created drought conditions in many parts of the country. The Jaipur district was one of the worst affected areas. It was estimated that about 73.64 percent of rural people in the district (2001) were in grip of severe drinking water famines[7]. This quantity was more in the previous drought years (1999, 2000) when less attention was paid towards the village drinking water programme by the government. In 2000, a large number of villages were hitted by severe drought due to scanty rainfall and dry-up wells from the close of the year; and drought conditions were accentuated from 2001. Sensing the impending danger to the human and cattle lives, relief work was started in the form of installation of handpumps, pumping from reservoirs, and tanker. The condition has created an acute shortage of drinking water. Summer is yet to set in but almost all the 1053 villages of the district are facing acute water shortage[8].

8.4. CAUSES OF WATER SCARCITY

The water table does not remain constant but rises and falls with recharge and depletion of ground water either due to rainfall or pumping out by tube-wells. The water table is shallowest, generally less than 15 m of fluctuations in water table occurs due to variation in ground water recharge, subsurface drainage and groundwater draft from the wells. Seasonal fluctuations of water level between pre-monsoon and post-monsoon period was observed by the C.G.W.B. during 1999-2001. In rainy season, the water table rises to just below the ground level and in summer it goes down beyond 11.85 metres at some places (Table 8.4).

Table 8.4 : Change in Water Level (Details of the Spot) (1994-2001).

No.	Village (Block)	Change in water level in (m)
1.	Bagru (Sanganer)	(-) 12.43
2.	Begus (Jhotwara)	(-) 12.52
3.	Chaura Rasta (Jaipur)	(-) 12.20
4.	Didwata (Phagi)	(-) 8.90
5.	Jamwa Ramgarh (J. Ramgarh)	(-) 8.45
6.	Kachnaria (Dudu)	(-)10.40
7.	Mandha (Viratnagar)	(-)8.81
8.	Sursinghpur (Dudu)	(-) 9.90
9.	Tolai (J. Ramgarh)	(-) 11.85

Source : State Remote Sensing Application Centre, Govt. of Rajasthan, Jodhpur.

To find out the trend of water table fluctuation, water table trends maps are drawn with the help of data obtained from observation wells. This shows very high water table fluctuation (above 20 meters) in area around Jhotwara, Govindgarh and east of Shahpura which has deep water level ranging more than 20 m. The average water table fluctuation in maximum area of the district has been in the range 10-20 meters southern-peripheral area and pockets scattered in different parts do not lie in this range.

Rain and its collected water are the main sources for irrigation and domestic purposes of southern and western part of the district. The tribal and backward people of this region are totally dependent on the vagaries of nature and so water supply condition is more acute in these areas.

The water level in Jaipur district has changed beyond 3 to 15 metres depth in 1994 to 2001. The average water level changes in southern and north-western peripheral region show marginal depletion in water level between the -3 to -7 metres range. Very high water level changes in major part of the district excluding southern peripheral region and pockets located in different part exhibits depletion in water level between the -7 to -15 range. In this area, high density of tubewells have been installed for irrigation purposes. During pumping, cone of depression is formed around the tubewells and water table goes down in the neighbouring wells. Water level of dugwells nearer to the pumping well goes down more than those further away. If they are not installed at a proper distance as in this area, the two or more cones of depression cut into each other and this creates general lowering of water table and all the wells in such area go dry. This conclusion is also in conformity "the area is under over-exploitations".

8.5 IMPACT OF YEAR 1999-2002 DROUGHT

This is the 4th year (2002) in succession of severe drought condition due to failure of monsoon. The severity of the drought in the current year is much more than in the past 3 years[9]. During the current monsoon period so far, none of the major water reservoir has either over-flown or received any significant quantity of water. Simultaneously, the water table has gone down alarmingly in these areas, causing an acute drinking water crisis in a large number of villages. Special arrangements were therefore made to transport drinking water by road and rail in addition to the efforts made by the Public Health Engineering Department through its water supply schemes[10].

Livestock play a vital role in the economy of the State since agricultural productions are less, by and large, due to erratic rainfall

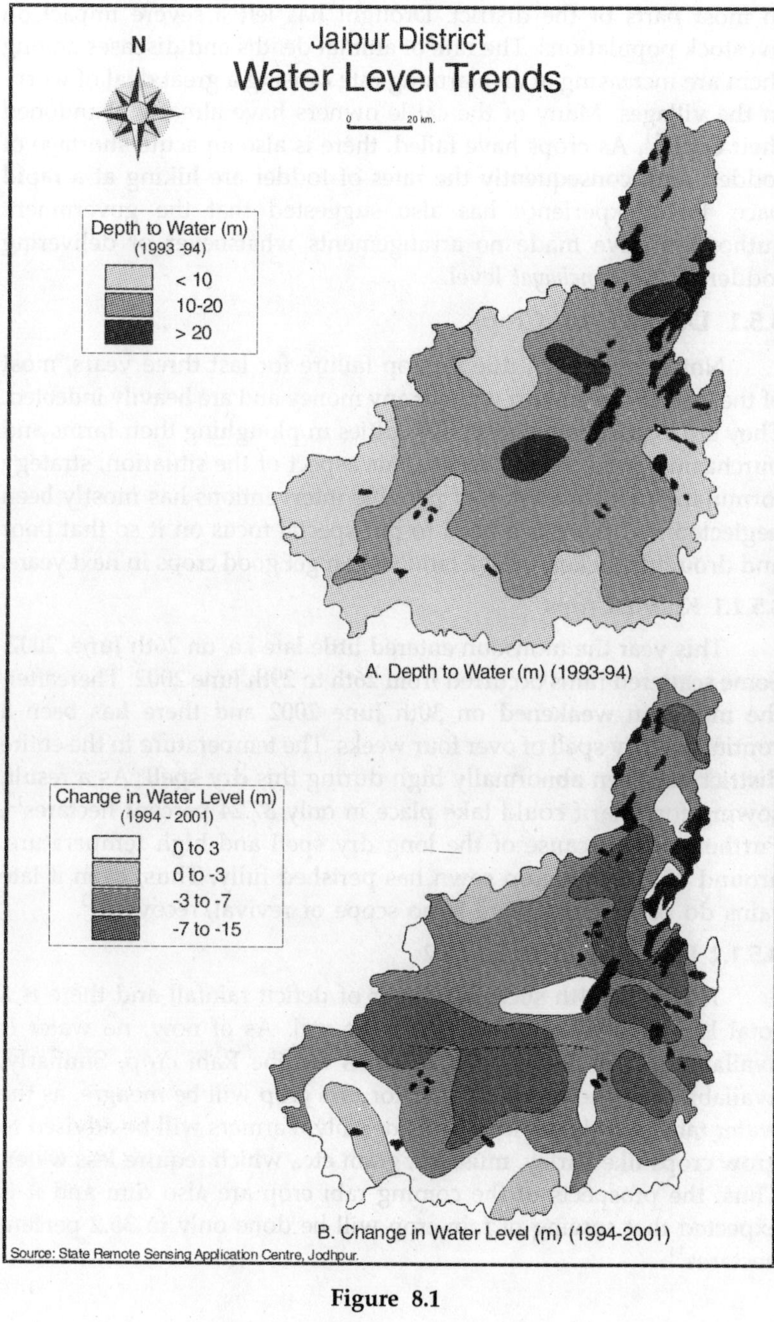

N
Jaipur District
Water Level Trends
0
20 km.
Depth to Water (m)
(1993 -94)
< 10
10-20
> 20
A. Depth to Water (m) (1993-94)
Change in Water Level (m)
(1994 - 2001)
0 to 3
0 to -3
-3 to -7
-7 to -15
B. Change in Water Level (m) (1994-2001)
Source: State Remote Sensing Application Centre, Jodhpur.

Figure 8.1

in most parts of the district. Drought has left a severe impact on livestock populations. The rate of animal deaths and diseases among them are increasing at an alarming rate causing a great deal of worry in the villages. Many of the cattle owners have already abandoned their cattle[11]. As crops have failed, there is also an acute shortage of fodder, and, consequently the rates of fodder are hiking at a rapid pace. Field experience has also suggested that the government authorities have made no arrangements whatsoever of delivering fodder at the *Panchayat* level.

8.5.1 Damage of Crops

Not surprisingly, due to crop failure for last three years, most of the farmers are almost without any money and are heavily indebted. They are experiencing great difficulties in ploughing their farms and purchasing good quality seeds. This aspect of the situation, strategy formulation and planning of possible interventions has mostly been neglected and there is a need to put special focus on it so that poor and drought stricken village families can get good crops in next years.

8.5.1.1 Kharif Crops

This year the monsoon entered little late i.e. on 26th June, 2002. Some scattered rains occurred from 26th to 29th June 2002. Thereafter, the monsoon weakened on 30th June 2002 and there has been a continuous dry spell of over four weeks. The temperature in the entire district has been abnormally high during this dry spell. As a result, sowing for Kharif could take place in only 37.24 percent hectares[12]. Further more because of the long dry spell and high temperature, around 80% of the crop sown has perished fully. Thus, even if late rains do occur now, there is no scope of revival/recovery[13].

8.5.1.2 Prospects of Rabi 2002

This is the 4th successive year of deficit rainfall and there is a total lack of residual moisture in the soil. As of now, no water is available in reservoirs/ dams/ tanks for the Rabi crop. Similarly, availability of the ground water for rabi crop will be meagre, as the water table stands depleted considerably. Farmers will be advised to grow crops like barley, mustard, gram etc., which require less water. Thus, the prospects of the coming rabi crop are also dim and it is expected that sowing of rabi crop will be done only in 35.2 percent hectares[14].

Women and children are the most affected groups in the drought. The lack of food, fodder and water in many areas is resulting in migration of men in search for employment. This leaves the womenfolk with increased burden to manage the household as well as trudge several kilometres to access drinking water. This affects the health and well being of the women and children as well[15].

Amongst the many adverse effects of drought, its ill effects on health have been a significant problem. During this crisis, it has been evident that pathological conditions such as malnutrition and deficiency syndromes have not been rare. Most commons of these have been iron deficiency Anemia, vitamin deficiencies (particularly Night blindness or Vitamin A deficiency) and malnutrition among children[16] There is a need of organising medical camps in most needy areas in which villagers can have a regular medical check up and could be provided necessary preventive and nutritive preparations.

8.5.2 Strategy to Combat Drought

The objective of drought relief activities will not be merely confining to providing sustenance to mitigate hardship but care for quality of life also. The State Government has planned the following strategy to combat the drought situation[17]:

- Wage employment to the people, particularly belonging to Below Poverty Line (BPL) families, on relief works.
- Under Relief works, the focus would be on asset creation with particular emphasis on water conservation, aforestation as a measure of long term drought proofing.
- Relief works would be dovetailed with State / Central sector schemes so that useful assets are created.
- Cattle conservation measures viz., supply of subsidised fodder, cattle camps for stray cattle, subsidy to goshalas, cattle feed subsidy, special arrangement for veterinary health care, mini-kits for growing green fodder, bull subsidy etc.
- Arrangements for drinking water in problematic areas.
- Active involvement of Non Government Organisation (NGOs), *Panchayat Raj* Institutions and donors in relief operations.
- Effective implementation of nutrition and health programmes.
- Gratuitous relief to destitute and infirm persons.

- Effective implementation of social security schemes.
- Food security through Food For Work programme.
- Transparency in relief works and social audit.
- 'Annapurna Scheme' : As per this scheme sacks of wheat should be available at all times in each *Panchayat* for free distribution to the destitute.
- The Govt. has decided to provide 20 kgs of wheat per family per month to the BPL families in villages. This distribution will be for free of cost.
- Repair of hand pumps : The PHED department of the state Govt. has decided to repair hand pumps in the villages of the State to improve water supply.
- Fodder depots: The Government has also decided to organise fodder depots in villages of State.
- Drinking water supply through tractor tankers in remote villages.
- Formation of Vigilance Committees: To check the effectiveness of various interventions; it has been planned by the Govt, to form such committees and post at least one person in each village.

8.5.3 Action Taken by The State Government

- Collectors have been directed to involve NGOs / non resident Rajasthanis and donors in relief activities. Meetings with NGO's and donors have been held by Collectors in all the districts.[18]
- In order to ensure transparency in the execution of relief works, order for compulsory social audit of relief works was issued on 17-08-2002. Social Audit would be carried out by the Gram Sabhas. A website – *www.rajgovt.org/relief.html* has been set up to provide all necessary information on relief activities.
- The Public Health Engineering Department has prepared a Contingency Plan of Rs. 502.90 crore to ensure the drinking water supply in the problematic areas and started its execution.
- In order to provide gratuitous relief to infirm and destitutes 5 quintals of wheat has been kept at each *Panchayat*

headquarter and 25 quintals at each tehsil headquarter to be distributed free of cost.

- Land Revenue collection has been suspended in all the districts.
- Short term cooperative loans are being converted into medium term loans.
- Revolving funds of Rs. 10 lacs for drinking water and Rs. 1.00 lacs for medicines both for human and cattle has been made available to each Collector to meet immediate and contingent needs.
- Contingency plan for shifting of cattle from drought effective districts to canal areas in case of need is being prepared.
- Special attention is being paid in the effective implementation of following schemes to save the poor people, pregnant mothers, old, children and infirm from malnutrition in the prevailing drought conditions[19].

(a) ICDS : The Scheme is under implementation in the entire Rajasthan State. The children of 0-6 year of age group and the pregnant mothers are being benefited through 35488 *Anganbaris.* At present 28.57 lac children are being benefited under this scheme.

(b) Mid-day Meal Programme : The scheme is being implemented for age group 6-11 years. The total number of beneficiaries under this scheme is 70 lac students.

(c) National Old Age Pension Scheme : Under National Old Age Pension Scheme 4,84,420 persons have been identified as potential beneficiaries. But Government of India is providing funds for 141496 persons at the rate of Rs. 75/- per month per person. Since under the State Pension Scheme, Rs. 200 per month are being given, the remaining Rs. 125/- is being given by State exchequer. In addition to this 219253 persons are being given pension @Rs. 200 per month under State Scheme. 197849 Widows are also being given pension in the State. 76866 disabled persons are also getting Disability Pension.

(d) Annapurna Scheme : Under the Annapurna Scheme, 1,05,293 persons have been selected and 1,04,469 persons are being benefited under the scheme. Under this scheme 10 kg food grains are provided free of cost to old persons of 60 years and above who are not beneficiaries under old age pension scheme.

(e) Antyodaya Ann Yojna : The scheme has been launched in the State from 4-3-2001. The poorest of the poor numbering about 3.72 lac families have been identified and issued special cards to avail the benefit under the scheme. 25 or 35 kg wheat @ Rs. 2 per kg is given to these families through public distribution system.

(f) Medicare cards for BPL Families : In addition to the above schemes, all the 23 lac BPL families in the State have been issued Medicare card which entitles them for free treatment including medicines in the Government hospitals. For serious illness, funds for free treatment are provided from "CM Life Saving Fund".

8.5.4 Weaknesses of Government Initiatives

The State Government's relief measures address only the immediate needs of the community and lack any long-term perspectives. The approach of the State Government to address drought is generally confined within the peripheries of service delivery (relief) rather then a long-term drought management programme. Main weaknesses of the programme are given below:

- The employment generation component of the relief operations could not go up to an adequate extent, and, therefore, a large contingent of drought affected families is still unemployed. Only few villagers are being benefited by the programme, and this is causing disharmony in the villages. (Table 8.5)

Table 8.5 : Labour Ceiling and Actual Labour Employed of Jaipur district (2001)

Population effected (in lacs)	Labour Ceiling	Actual labour employed	Percentage of labour employed with reference to population	Percentage of labour employed with reference to Ceiling
23.42	11000	11305	0.482	102.77

Source : Health Environment and Development Consortium, Jaipur.

- Most of the work done under the Food for the Work programme, is still concentrating on construction of roads etc., rather than building up community water harvesting water structures for long term drought proofing.
- Annapurna Scheme has not been a success; as per the table

8.6 only 20 percent (2266 MT.) wheat is distributed in Jaipur district, since the distribution system has not been very effective[20].

Table 8.6 : Lifting of Wheat in Relief Works (Jaipur district-2001)

(in MT)

Quantity Allocated	Quantity Lifted	Quantity Distributed	Balance
11293.00	8514.00	2266.00	9027.00

Source : *http://www.rajgovt.org/relief/*, Official Website Portal of Govt. of Rajasthan.

- The Government's decision of providing 20 kgs of wheat to BPL families does not sound very sensible since the average size of a family is 6 to 7 people. Also, the Government figures of the BPL families is very old and the recurrent drought have converted the Above Poverty Line (APL) status of most rural families in BPL status leaving a large number of unofficial and unidentified BPL families (which are APL according to Govt. Statistics).
- The same thing could be said for the Food for Work Programme also. Government guidelines suggest that only BPL families should be given the benefit of the programme which leaves the other families unbenefited by the programme.
- Govt. of Rajasthan has sanctioned 2185 relief works for Jaipur district in year 2001, but only 1416 works have started (Table 8.7) as yet.

Table 8.7 : Progress of Relief Works, Jaipur District (2001)

No. of works sanctioned	No. of works started	Labour ceiling	No. of Labour engaged	Average wage paid per labour
2185	1416	31737	25169	55.00

Source : *http://www.rajgovt.org/relief/*, Official Website Portal of Govt. of Rajasthan.

- The repair of hand pumps by the PHED department is a good idea but unfortunately, not much has been done on this front.

- The supply of drinking water is also not adequate and being influenced by the rich and powerful sections of villages.
- Fodder depots have been set up only at a few places.
- Vigilance committees formed by the Government in the villages are not very active. It has been reported that in most of the villages, they are non-existing and non-operational.

People of the State are starving of hunger and foodgrains are lying in Food Corporation of India (FCI) godowns. Government of India (GOI) has allocated 1 lac M.T. of wheat for Food for Work programme to the Rajasthan State. The allocation had many conditions (like Govt. of Rajasthan can use the foodgrains for below powerty line BPL families and under the programme not more than 5 kg of wheat can be given to each labourer employed under food for work) which made it difficult for Govt. of Rajasthan (GOR) to use it properly. All the APL masses due to third consecutive drought have fallen to BPLs, and GOI restrictions are not allowing them to work under the schemes. But GOR did not followed the GOI stipulation and provided the work to all levels according to the ceiling of the work. On March 5, 2001, NGOs and villagers in several districts organised a demonstration outside the FCI godowns to draw the attention of the people and the government to the ironical fact that 50 million tonnes of foodgrains was rotting in FCI godowns even as people were dying from starvation[21].

During kharif season most part of the State received less and late rainfall which led to damage of 50% crop. Furthermore less availability of water in tanks and wells led to failure of winter crop also. Many agitation were staged by farmer organizations and voluntary organisations demanding atleast 8 hours of electricity for irrigation purposes. Government of Rajasthan promised many times but failed to supply 8 hours of electricity, which resulted into huge losses of rabi crops. Estimated loss is about 1847 crores. On Feb 19, 2001 at many places in Rajasthan, farmers protested against Electricity Department for supply of adequate supply of power to fetch water for irrigation. Replying to the debate in the State Assembly, Hon'ble Chief Minister, GOR said "I understand how farmers feel when they do not get electricity for irrigation. We have been making all out efforts to supply eight hours of electricity to farmers but we are dependent on our neighbours for additional electricity"[22].

There is very less coverage on livestock as most of the people have left their cattle on God, as they themselves are starving for food and water. Almost all variety of cattle, wild animals and birds has been hit by fatal and chronic diseases this year[23]. Due to thirst and dehydration 11 peacock died in Jaipur district on 29 April 2001[24]. With scarcity of fodder the milk production has gone down[25]. Although some rotten foodgrain (wheat and rice) has been allocated as cattle feed by GOI[26] but proper distribution is still to be done. During last year's drought GRAVIS opened about 17 fodder depots of which 6 were supported by GOR, but till now no payment has been made by administration to them. There are many such case-reports because of which volunteers have lost their confidence in opening fodder depots for government.

The Government's failure to provide adequate drought relief is a violation of the affected people's constitutional rights, including the right to live. It is also an economic blunder. Indeed, unrelieved drought leads to a costly depletion of productive capital as drought affected people lose their cattle, starve, run out of seeds for the next crop, and degrade the local environment in search of subsistence. The response of government at all levels is that they have no money. But this refrain does not cut much ice when 50 million tonnes of foodgrain are lying idle in the godowns of FCI. The sight is an injustice to the hungry masses of the State.

REFERENCES

1. Rathore, M.S. (2002) : "Drought Management in Rajasthan: Some Long Term Policy Issue", *Drought issue in Rajsthan,* Rajasthan State Public Administration Institute, Jaipur.
2. Website : *http://www.rajgovt.org/relief2002/,* Official web Portal of Govt. of Rajasthan.
3. Website : *http://www.rajirrigation.com,* Irrigation Department, Govt. of Rajasthan.
4. Surbrahmanyam, V.P. (1975) : "Water Balance Technique in Drought Climateology", *Vayu Mandal,* Vol.5, No.4, p. 75.
5. GOI (2001) : *Monsoon 2001,* Irrigation Department, Govt. of Rajasthan, Jaipur. p.6
6. GOI (2001) : *Ibid,* 7
7. Website : *http://www.org.in/undmt/states/rajas,* United Nations Disaster Management Team.

8. Newspaper (2001) : *Hindustan Times,* dated 6 March, 2001.
9. GOR (2002) : *Interim Memorandum on Scarcity (Samvat-2059),* Relief Department, Govt. of Rajasthan. p. 1
10. GOR (2002) : *Ibid,* p. 2
11. Malik, D. (2001) : *Drought In Rajasthan Current Status of Situation and Interventions,* Health Environment and Development Consortium. p 5.
12. GOR (2002) : *Interim Memorandum on Scarcity (Samvat-2059),* Relief Department, Govt. of Rajasthan. p. 3
13. GOR (2002) : *Ibid,* p. 3
14. GOR (2002) : *Ibid,* p. 4
15. Website : *http://www.rajgovt.org,* Official Web Portal of Govt. of Rajasthan.
16. Malik, D (2001) : *Drought In Rajasthan Current Status of Situation and Interventions,* Health Environment and Development Consortium. p. 6
17. GOR (2002) : *A Note on Drought Situation in The State and Its Management (Samvat 2059),* Relief Department, Govt. of Rajasthan, p.4-5.
18. GOR (2002) : *Ibid,* pp. 6-7
19. GOR (2002) : *Ibid,* pp. 10-11
20. Zdres, J. (2001) : *Hunger Admist Plenty,* A Few Facts & Figures (Oct to Dec 2000)
21. Newspaper (2001) : *Hindustan Times,* dated 6 March, 2001.
22. Newspaper (2001) : *Hindustan Times,* dated 19 April, 2001.
23. Newspaper (2001) : *Hindustan Times,* dated 30 April, 2001.
24. Newspaper (2001) : *Dainik Bhasker,* dated 26 Feb. 2001.
25. Newspaper (2001) : *Hindustan Times,* dated 6 March, 2001.
26. Website : *http://www.rajgovt.org/relief/,* Official web Portal of Govt. of Rajasthan.

9

Planning the Rural Water Supply

Provision of safe and improved drinking water supply is one of the principle elements as well as a leading step towards primary health care. Access to safe water improves the quality of life within entire community and frees mothers from the burdensome task of carrying water over long distances. Moreover, the provision of village water supplies can be taken as catalytic element for a widening circle of health orientation and other co-operation, self-help, community efforts such as sanitation activities, food production and local support of health and education services. Such efforts can contribute greatly to the delivery of simple preventive health care by giving community the capacity to maintain a healthy environment for all its members.

9.1 HISTORICAL OUTLINE

Right from the time of Independence both the Governments of the State of Rajasthan and India have taken keen interest to implement the drinking water supply scheme for rural population. However, the achievements could not be significant because of financial constraints. During the Eighth and Ninth Five Year Plans, 1539 villages in Jaipur district were modernised with water supply facility and extended handpump schemes to pipe or regional schemes[1], but these schemes ceased to function for want of funds for proper and effective maintenance.

In the approach paper on the Ninth Five Year Plan, emphasis has been laid to provide minimum basic needs (MBN) on highest priority. Provision of safe and adequate quantity of drinking water is given priority under this scheme. All the States have been directed by the Government of India to draw up detailed plans for providing drinking water supply to all the villages by the end of Ninth Plan.

In Rajasthan, Public Health Engineering Department, Government of Rajasthan has conducted survey and investigations of the State and has concluded that 1585 out of total of 2131 inhabited villages in the Jaipur district, experience acute scarcity of water[2]. These scarcity-villages occur in almost all the Jaipur district.

9.2 CURRENT APPROACHES

Rural water supply was one of the very important items of Ninth Five Year Plan (Figure 9.1)[3] and Action Plan (1999-2000) (Figure 9.2)[4] of Public Health Engineering Department. It was decided to provide safe and adequate potable water to all the problem villages including non-scarcity villages also. Although it was decided to provide water through pipe line, but funds for construction was the main problem than its maintenance. So, at a first step, water supply scheme was prepared on the basis of collection of some water tax from local population. High population concentration will increase shortage of water and problem of health (Table 9.1).

NINTH PLAN PROVISIONS

URBAN	**795 Cr.**
RURAL	**965 Cr.**
	(Including 164 Cr. of German project & 1 Cr. for ESTI)

- To cover about 22000 habitations and 3000 pc to fc habitations.
- Strengthening of urban and rural water supply schemes.
- Implementation of schemes of excess fluoride and salinity areas
- Modernisation of existing water supply schemes.
- Implementation of sewerage schemes.
- Implementation of german assisted churu project.

Source : PHED, Jaipur (1998)

Figure : 9.1

Table 9.1 : Characteristics of Settlement and Water Supply

Pattern of Rural Settlement	Type of Sources Available	Level of Health Hazard From Water	Cost Level (energy/cash)
Scattered	Surface-pond, tanks, lakes, stream	High	High
	Underground -spring wells	Low	Low
Clustered	Taps	Low	High
	Stand pipes	Low	Medium
	Surface-pond, tanks, lakes, stream	High	Low
	Underground -spring wells	High	Low
	Shallow wells, deep wells	Low	Low
Urban	Taps	Low	High
	Stand pipes	Medium	Medium
	Surface-pond, tanks, lakes, stream	High	Low
	Underground-spring wells	High	Low

The distributional pattern of rural settlement varies from one to another area and it is not feasible to connect all the areas through pipe line. After 30 years of starting water supply scheme, most of the villages are left out without any consideration. For these villages, some programme should be launched, e.g. construction of hand-pumps, sanitary well, protection and maintenance of old drinking wells, and help in maintenance of wells (to be provided by the local authorities). The Rural Piped Water Supply Programme was started as early as 1972 but the villages actually receiving water through these schemes are not more than 122 (only 5.73 percent) in the whole district. Although in the Eighth and Ninth Five Year Plans, there is big amount of money available for the work, but there is some major defect in release of money (e.g., release of funds in instalments which creates hindrance in continuous work) and work is not done continuously as per plan; hence the pipe water supply scheme is not as much-successful as it should have been.

9.3 PLANNING AND PROGRAMME FOR FUTURE NEED

Water resource development programmes should, as far as possible, be planned and developed as multipurpose use. Provision for drinking water should be a primary consideration. The programme should provide water for irrigation, food mitigation, hydro-electric power generation, navigation and recreation wherever possible[5].

Action Plan Year 1999-2000

- Completion of incomplete works of rural and urban water supply schemes.
- augmentation of existing schemes and improving water pressure in towns.
- reducing wastage of water.
- coverage of 4500 nc habitations & 600 pc to fc habitations.
- conversion of hand pump schemes to p&t (population > 1500) & piped schemes (population > 4000) to deal with problem of quality/failure of sources.
- ensuring scheme give full benefit.
- compliance of the announcement made by hon'ble minister phed on floor of the house.
- implementation of the schemes of excess flouride and salinity areas.

Maintenance

- staff should be present and responsive.
- adequate touring by officers should be ensured.
- low pressure areas in urban towns to be attended at all levels.
- water reaching tail end villages should be ensured.
- leakages should be removed.
- maintenance of piped, p&t and regional schemes will be on regular basis.
- hand pumps in interior areas will be carefully attended to.

***Source** : PHED, Jaipur (1999)*

Figure : 9.2

9.3.1 Bisalpur Project Nearby Jaipur District

The Bisalpur Project is a multipurpose project i.e. irrigation cum drinking water supply project. It envisages construction of a storage dam across river Banas near village Bisalpur in Tonk district to provide drinking water supply to Jaipur, Ajmer, Beawer, Kishangarh, Nasirabad and other enroute cities, towns and villages. It's has been designed with gross capacity of 1095.84 Mcum (Table 9.2).

Table 9.2 : Salient Feature of Bisalpur Dam

No.	Feature	Measurement
1	Location	On Banas near village Bisalpur in Deoli Tehsil, Tonk District (Rajasthan)
2	Catchment Area	
	(a) Gross	27725 Sq. Km. (10705 Sq. Miles)
	(b) Intercepted	19071 Sq. Km. (7363 Sq. Miles)
	(c) Free	8655 Sq. Km. (3342 Sq. Miles)
3	Average Annual Rainfall	588 mm (23.15 Inches)
4	Storage Capacity	
	(a) Gross	1095.84 M cum (38700 Mcft)
	(b) Live	893.91 M cum (31568 Mcft)
	(c) Upto Crest level	22.43 M cum (792 Mcft)
5	Annual Yield	
	(a) Yield at 75% Dependability	938.73 M cum (33150 Mcft)
	(b) Yield at 50% Dependability	2432.45 M cum (85900 Mcft)
6	Peak Design Floor Discharge	33800 cumecs (1.19 M cusec)
7	Important Levels (Above M.S.L.)	
	(i) Top of Dam	EL 322.50 m
	(ii) Top of Pier	EL 324.50 m
	(iii) Maximum Water Level	EL 316.35 m
	(iv) Full Reservoir Level	EL 315.50 m
	(v) Spillway Crest Level	EL 302.30 m
8	Length of the Dam	574 m (1883 ft)
9	Length of Non Overflow	236 m (774 ft)

Contd...

...Contd.

No.	Feature	Measurement
10	Length of Overflow	338 m (1109 fl)
11	Maximum Height of Dam	
	(a) Above river bed	27.5 m (90 ft)
	(b) Above deepest foundation	39.5 m (130 fl)
12	Maximum Design discharge	29047 cumec (103 Mcuse)
13	No. of crest Gates & Size	18 gates of 15 mi x 14 m
14	Reservation of Water (Excluding reservoir Losses)	
	(a) Drinking Water	458.46 M cum (16200 Mcft)
	(b) Irrigation	226.40 M cum (8000 Mcft)

Source : Irrigation Department, Govt. of Rajasthan, Jaipur

The project was initially approved by Central Water Commission, New Delhi dated 28 April 1993 amounting to Rs.309.07 crores at 1991 price level, now the revised estimate cost of Rs.657.91 crores (March 2000 price index[6].

Bisalpur Dam site drains the areas of Bhilwara, Udaipur and Chittorgarh district where average rainfall is comparatively less than the average rainfall of the areas on down stream of Bisalpur Dam, through maximum rainfall is higher in upstream area of Bisalpur than downstream area. This indicates that if a suitable site can be located on river Banas, down stream of Bisalpur Dam, there is a possibility of obtaining additional water for meeting the requirement for Jaipur district particularly when Bisalpur Dam receives inflow much less than the average inflow.

Due to the submergence, land and properties of 63 villages have been affected out of which 32 villages belong to Deoli Tehsil, 13 villages to Todaraising Tehsil, 17 villages to Kekri Tehsil and one village to Jahajpur Tehsil. The total population affected is about 30,000 (1991 census) out of which 5900 persons belong to scheduled caste, 5700 persons to scheduled tribe and 18400 belong to other caste. The total numbers of affected families are 5700[7]. As per norms approved by Govt. of Rajasthan, compensation of land and properties are to be paid according to level of submergence in Stages, by considering villages as a unit in accordance with Central Land Acquisition Act 1884 and as per decision taken in the meeting chaired by Chief

Secretary, Govt. of Rajasthan on 17.11.94. The entire compensation of land and property, total amounting to Rs.121.00 crores has already been paid upto March, 2002. The Status of various components of the project is as under[8]:

9.3.1.1 Dam Alongwith Allied Works

Dam alongwith intake structure for water supply to Jaipur and Ajmer districts has already been completed and installation of gates was done in the year 1997-98. Partial storage is being done in the dam since 1994 for providing drinking water to Ajmer, Kishangarh, Beawar and enroute villages. Water in the reservoir was stored upto RL 311 M during the year 2001 which is being utilised mainly for drinking purpose as per requirement of PHED and for irrigation. A potential of 18,450 ha. has been created in this project.

9.3.1.2 Canal Works

There are two canals namely Right Main Canal (RMC) and Left Main Canal (LMC) each of length 51.64 Km. and 19.00 Km respectively (Table 9.3). Construction of RMC has been completed in 45 Km. length and its distribution system is ready upto this length and therefore potential has been created for 18450 hectares of land. Construction work is in progress between 46 Km. to 51.64 Kms length of RMC. Similarly, work is under progress in the reach 0 to 4.5 Km of LMC.

Table 9.3 : Canals Feature of Bisalpur Dam

No.	Name of Canal	Total length of canal in k.m.	Completed canal length in k.m. (on March 2002)	Command area in hectares	No. of Villages to be benefited
1	Right Main Canal	51.64	45.00	69393	218
2	Left Main Canal	19.00	0	12407	38
	Total	**70.64**	**45.00**	**81800**	**256**

Source : Irrigation Department, Govt. of Rajasthan, Jaipur

9.3.2 Proposed Bisalpur Jaipur Pipeline

The present water supply arrangement for Jaipur is primarily dependent on groundwater. However, groundwater situation in Jaipur is going to be very critical in 4-5 years both in terms of quantity and quality. As an alternate source of water, Bisalpur Dam is being envisaged. The project would involve water treatment and transmission by about 105 Kms., with bulk transfer within Jaipur at various points.

Bisalpur Dam has already been designed and is under construction with capacity of 893.91 Mcum (Live Storage) (Table 9.2), out of which 460 Mcum have been reserved for drinking water needs of Jaipur, Ajmer and enroute villages and towns along the pipe line route[9].

Different water supply schemes of Jaipur district which can be linked with Bisalpur water supply system, a pipe line carrying the requirement of water sufficient to meet the future needs, can be designed to link the off take point from the Bisalpur water supply system to the clear water reservoirs of Rural Water Supply Schemes (RWSS) presently collecting water from the tubewells. At present the pipe network for the RWSS is being designed for a 20 LPCD supply but in future it is expected that power will be available at the source village for 16 hours in place of 8 hours and then it should be possible to supply 40 LPCD from the existing pipe network.

The villages which can be served by Bisalpur dam have been grouped separately and the requirement of water for scarcity rural areas which can be drawn from Bisalpur dam are shown in Table 9.4. While designing the storage capacity of Bisalpur Dam, the horizon year is kept as 2021. Therefore, future water demand of scarcity village for the years 2011 and 2021 should be met from Bisalpur Dam in addition to the water demand of enroute villages.

The requirement of water beyond the year 2010 for all the scarcity villages of the district as well as for the villages having other quality problems i.e. presence of excess fluoride, nitrates and TDS has been assessed by considering the extreme limits of WHO standards prescribed for Rajasthan (Appendix-4). The demand on Bisalpur Dam will thus be an additional demand of water (Table 9.5).

The following items have been considered to work out total water demand from Bisalpur Dam.

- Where the present per capita water supply is less than 40 LPCD and which are in proximity of the Bisalpur pipe line alignment after deducting the quantity of water at present service level.
- Scarcity villages for which water supply schemes have been designed and which are in proximity of the Bisalpur pipe line alignment. Thus, the total water requirement that should be met from Bisalpur Dam in the 2011 and 2021 has been worked out and is shown in Table 9.4.

Table 9.4 : Projected Population and Water Demand for Rural Areas which can be connected by Bisalpur Water Pipe Line (2001)

No.	Name of Scheme	Population (2001)	Projected Population 2011	2021	2031	2041	Water Demand (MLD) 2001	2011	2021	2031	2041
1	Bindayaka-Neemera	1616	1778	1938	2093	2198	0.06	0.07	0.08	0.08	0.09
2	Laxminarayanpura-Bankrota	14989	16488	17972	19410	20380	0.60	0.66	0.72	0.78	0.82
3	Ramsinghpura-Ganpathpura	481	529	577	623	654	0.02	0.02	0.02	0.02	0.03
4	Chaksherwali-Chattarpura	237	261	285	307	323	0.01	0.01	0.01	0.01	0.01
5	Shrirampura-Ramsinghpura	9108	10019	10920	11794	12384	0.34	0.40	0.44	0.47	0.5
6	Gunwar Brahmin-Kheri Gokulpura	4796	5276	5751	6211	6522	0.19	0.21	0.23	0.25	0.26
7	Khoosar-Srikishanpura	7506	8257	9000	9720	10206	0.30	0.33	0.36	0.39	0.41
8	Saligrampura-Teerota	4344	4778	5209	5625	5906	0.17	0.19	0.21	0.22	0.24
9	Looniyawas-Hirapura	864	950	1035	1118	1174	0.03	0.04	0.04	0.04	0.05
10	Balapura-Mundor	11861	13047	14222	15359	16127	0.47	0.52	0.57	0.61	0.65

Contd...

...Contd.

No.	Name of Scheme	Population (2001)	Projected Population 2011	2021	2031	2041	Water Demand (MLD) 2001	2011	2021	2031	2041
11	Mandaliyajoga - Jaichand Ka Bas	755	831	905	978	1027	0.03	0.03	0.04	0.04	0.04
12	Kagya-Bhogpura	1085	1193	1301	1405	1475	0.04	0.05	0.05	0.06	0.06
13	Narikhera-Jaisinghpura	8086	8894	9695	10470	10994	0.32	0.36	0.39	0.42	0.44
14	Mardawara-Kudli	4657	5123	5584	6031	6333	0.19	0.20	0.22	0.24	0.25
15	Lakhawas-Sameliya	1346	1481	1614	1743	1830	0.05	0.06	0.06	0.07	0.07
16	Datooli-Jharana	330	363	395	427	448	0.01	0.01	0.02	0.02	0.02
17	Ganeshpura-Pahariya	1222	1345	1466	1583	1662	0.05	0.05	0.06	0.06	0.07
18	Shrijaipalpura-Decomand	1674	1842	2008	2168	2277	0.07	0.07	0.08	0.09	0.09
19	Dosra Khurd - Ramsinghpura	703	774	843	911	956	0.03	0.03	0.03	0.04	0.04
20	Muratpura-Bisaloo	970	1067	1163	1256	1318	0.04	0.04	0.05	0.05	0.05
21	Rathanpura-Nathmalpura	371	408	445	481	505	0.01	0.02	0.02	0.02	0.02
	Total	77001	84704	92328	99713	104699	3.03	3.37	3.70	3.98	4.21

Source : Irrigation Department, Govt. of Rajasthan, Jaipur.

Table 9.5 : Additional Requirement of Water Which Can be met from Bisalpur Dam

No.	Description	Water Demand (MLD) 2011	2021
1	Domestic Use		
	a. Urban Water Supply	38.28	52.45
	b. Rural Water Supply	3.39	3.69
2	Industrial Use	7.66	10.49
3	Miscellaneous	12.33	16.66
4	Losses in Storage and Transmission	15.41	20.82
	Total Water Demand	77.07	104.11
	Annual Water Demand (in ML)	28131.53	37999.16

This requirement of water is proposed to be assessed for villages, Dhanies and Magras located in the districts @ 70 litres per capita per day. The assessment of requirement of water beyond the year 2010 is being done for taking advance action for reserving the allocation of water from a new source of water.

It may not be possible to cover all the scarcity areas of the district from a new surface source of water. For this purpose only such villages and towns have been selected which are near the alignment of the pipe line proposed from Bisalpur dam to supply water for Jaipur. For other towns and villages, in future, additional sources of water will have to be found out which may consist of either creating storage of fresh water in subsoil or constructing a new impounding surface reservoir.

9.3.3 Proposed Dam at Isarda Water Reservoir

In addition to some villages and towns which though do not fall enroute of Bisalpur-Jaipur pipeline yet where the service levels are less than 40 LPCD; and for which the existing ground water sources are inadequate, a new surface source should be selected and a sufficient provision should be made in the storage reservoir while designing the project to meet the water demand of 2031-2041. The proposed Isarda Dam project should be considered for this purpose.

On the basis of hydrogeological studies made by the Irrigation

Department additional sources of water has been suggested by constructing a dam on river Banas near Isarda village. From this dam the future needs of drinking water of Jaipur district can be met with. Because the drinking water demand will be building up slowly, this water will have to be reserved for future drawl. It is, therefore, recommended that suitable legislation should be enacted to reserve the available quantity of water for drinking purposes and a suitable institutional arrangement should be established for implementation of the provisions made in the new legislation. This led to a proposal that utilisation of Bisalpur and Isarda dams should be taken as composite project and water available should be utilised rationally for irrigation and drinking purposes.

Irrigation Department construction of a dam on Banas river near village Isarda to utilise the run-off from the catchments area between Bisalpur and Isarda @ 95 percent dependability, is estimated at 160 Mcum. A preliminary project has also been prepared by PHED, to supply drinking water to Jaipur and to enroute villages from Isarda dam.

In view of the prevailing circumstances, there can be more than one options for supplying drinking water to Jaipur:

- Supplying entire quantity of 11100 Mcft. from Bisalpur dam and for future needs to draw 5500 Mcft. of water from Isarda Dam.
- Supplying 5500 Mcft. of water from Bisalpur dam and approximately 11000 Mcft. of water from Isarda dam.
- Supplying 5500 Mcft. of water from Bisalpur dam and 5500 Mcft. of water from Isarda and remaining to be utilised for Irrigation.

9.4 ISSUE AFFECT WATER USE PATTERN

The volume of water used by people who carry it home seems to be associated with the number of people in the household in a variety of cultures. If two wells are situated at an unequal distance from a house having bad quality water in nearer one and good quality in the well situated at longer distance, the households using bad water (hard water) for cooking, washing utensils and clothes, and for bathing, due to distance factor for saving time and energy, creates some problems in both cooking and washing. The well situated

Water Supply Choice and Use Model for Low-Income Communities

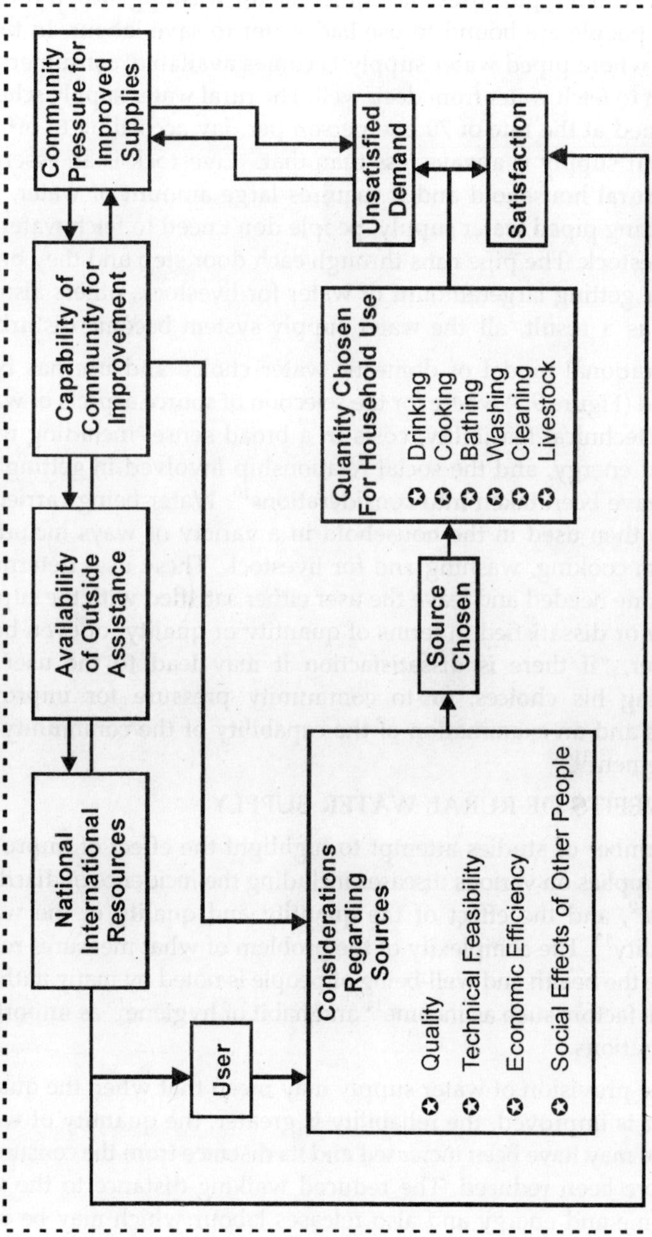

* (Adopted from White et al., 1972)

Figure : 9.3

comparatively at longer distance having good quality water is used only for drinking purposes and sometime for washing also. It reveals that the people are bound to use bad water to save labour. In those villages where piped water supply becomes available, consumers do not want to fetch water from deep well. The rural water supply scheme is designed at the rate of 70 one person per day according to official report but supply is always less than that. Livestock is an essential part of rural household and it requires large amount of water. But after getting piped water supply, people don't need to fetch water for even livestock. The pipe runs through each door step and they break them for getting largeramount of water for livestock, others also do so and as a result, all the water supply system become disturbed.

A rational model of domestic water choice and use has been proposed (Figure 9.3) where for the selection of source aspects of water quality, technical feasibility, costs in a broad sense, including time, cash and energy, and the social relationship involved in getting the water, have been taken into considerations[10]. Water being carried at home is then used in the household in a variety of ways including drinking, cooking, washing and for livestock. These uses determine the volume needed and leave the user either satisfied with the supply of water or dissatisfied in terms of quantity or quality, or even both. Moreover, "if there is dissatisfaction it may lead to the user re-examining his choices, or to community pressure for improved supplies and an examination of the capability of the community for improvement[11]"

9.5 BENEFITS OF RURAL WATER SUPPLY

Number of studies attempt to highlight the effect of improved water supplies on various diseases including the incidence of diarrheal diseases[12], and the effect of the quantity and quality of the water availability[13]. The complexity of the problem of what measures really improve the health and well-being of people is noted by many authors, who cite factors such as income[14] and habit of hygiene[15] as important considerations.

The provision of water supply may mean that when the quality of water is improved, the reliability is greater, the quantity of water available may have been increased and its distance from the consumers may have been reduced. The reduced walking-distance to the well saves time and energy and also releases labour which may be used for productive purposes leading to an economic benefit. Even if the economic value of the saved labour is zero there is a social benefit from the time saved and drudgery eliminated because more time can be spent with the family or on domestic activities and leisure. The improved quality of the water may lead to improved health which can result in

- lower medical expenditure,
- an improved sense of well-being,
- increased fitness and hence increased productivity of agricultural workers leading to an increase in crop production, and
- a lessening of the debilitating effects of childhood infections. The reliability of the supply will reduce the risk of epidemics.

If water is available for cattle near the homestead, the condition of the animals may improve; if they do not have to walk so far and are exposed to fewer diseases, the people may then be able and willing to invest in better quality of cattle. The potential benefits from water supply scheme may be illustrated schematically (Figure 9.4).

Water-borne disease such as dysentery can be prevented by improving the quality of water, e.g., by chlorination. Water-washed diseases such as skin-infections are associated with the need to wash. They can be significantly decreased by using increased quantities of water. The potential benefits will only be achieved if people use the hygienic water supply rather than their traditional source. To enforce the consumers to use an improved supply, it should be atleast as near to their home as their natural source. If improved supplies are distant, people only use the supply for part of the year, and in other areas some people may not use it at all.

An important drawback of improved water supply is that it reduces the resistance towards the water related diseases. Some experts comment that a very occasional breakdown could be serious, if by using good quality water for a long time people lose some of their present day immunity to the more common sickness, and it may also occur when an improved water supply area people visit their relations in other villages, where such improved water is not available.

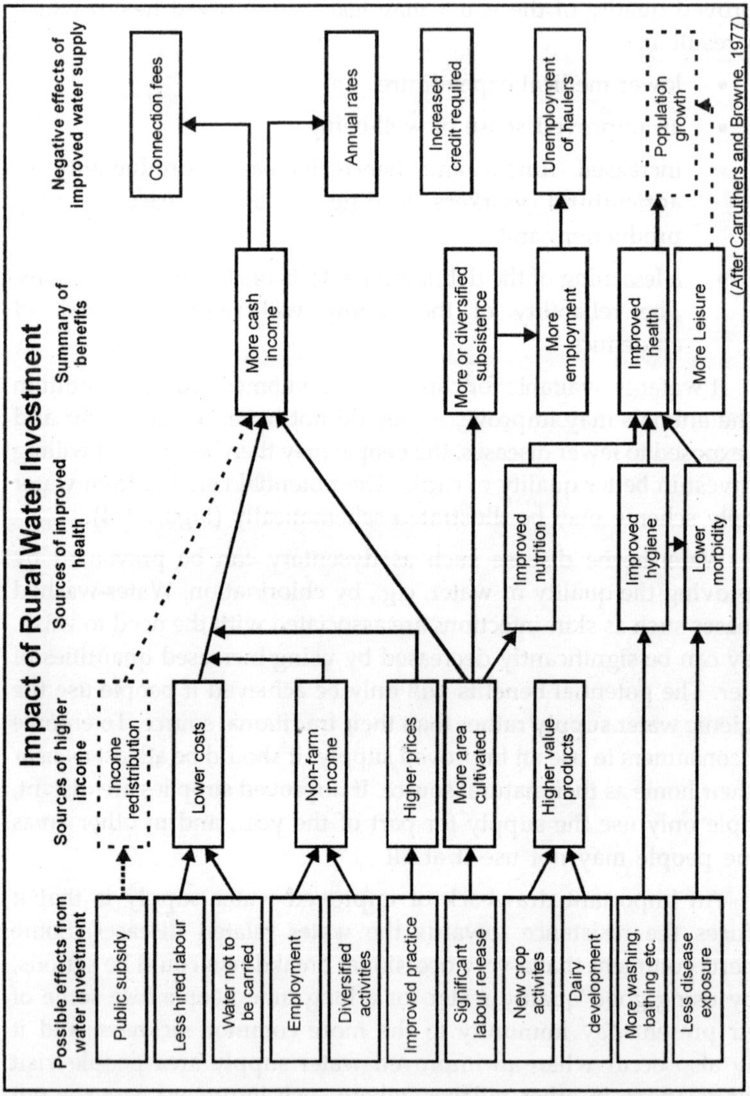
Impact of Rural Water Investment
Possible effects from water investment
Sources of higher income
Sources of improved health
Summary of benefits
Negative effects of improved water supply
Public subsidy
Less hired labour
Water not to be carried
Employment
Diversified activities
Improved practice
Significant labour release
New crop activites
Dairy development
More washing, bathing etc.
Less disease exposure
Income redistribution
Lower costs
Non-farm income
Higher prices
More area cultivated
Higher value products
Improved nutrition
Improved hygiene
Lower morbidity
More cash income
More or diversified subsistence
More employment
Improved health
More Leisure
Connection fees
Annual rates
Increased credit required
Unemployment of haulers
Population growth
(After Carruthers and Browne, 1977)

Figure : 9.4

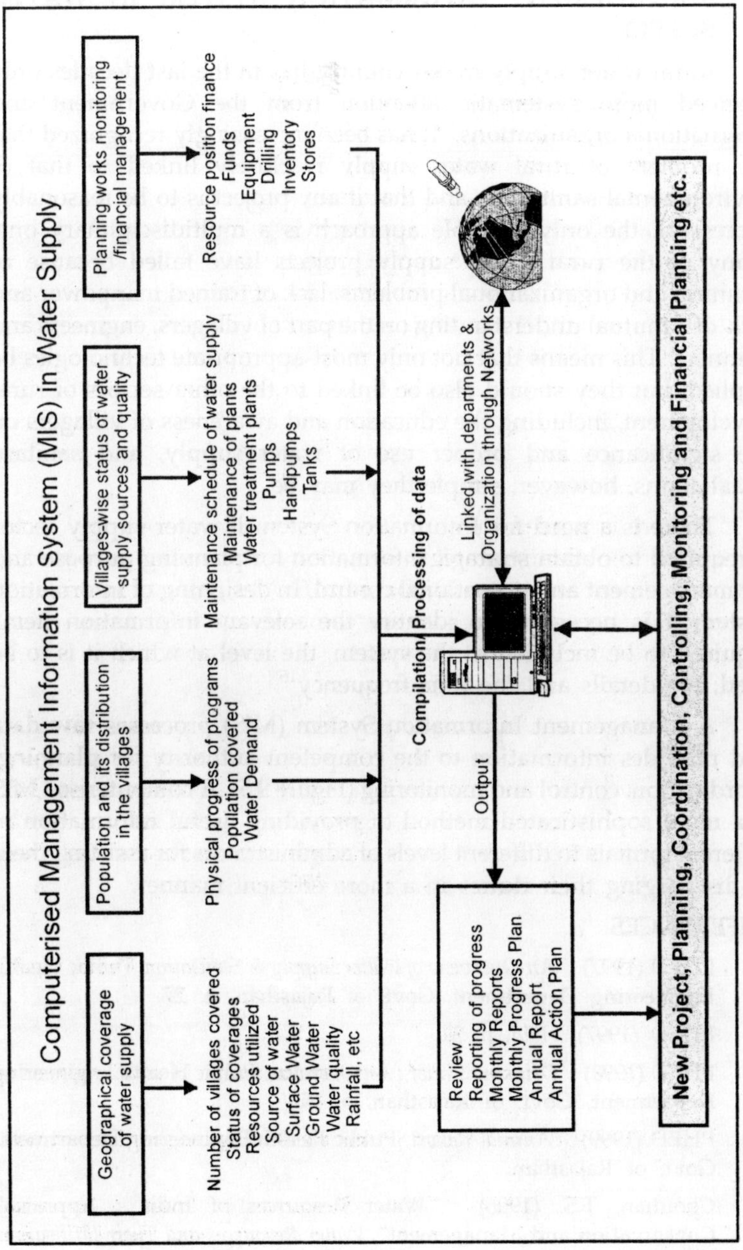
Computerised Management Information System (MIS) in Water Supply
Geographical coverage of water supply
Population and its distribution in the villages
Villages-wise status of water supply sources and quality
Planning works monitoring /financial management
Number of villages covered
Status of coverages
Resources utilized
Source of water
Surface Water
Ground Water
Water quality
Rainfall, etc
Physical progress of programs
Population covered
Water Demand
Maintenance schedule of water supply
Maintenance of plants
Water treatment plants
Pumps
Handpumps
Tanks
Resource position finance
Funds
Equipment
Drilling
Inventory
Stores
Compilation/ Processing of data
Linked with depaments &
Organization through Networks
Output
Review
Reporting of progress
Monthly Reports
Monthly Progress Plan
Annual Report
Annual Action Plan
New Project, Planning, Coordination, Controlling, Monitoring and Financial Planning etc.

Figure : 9.3

9.6 MANAGEMENT INFORMATION SYSTEM IN WATER SUPPLY

Rural water supply in our country has in the last decades only received more systematic attention from the Government and International organizations. It has been increasingly recognized that the problem of rural water supply is closely linked to that of environmental sanitation, and that if any project is to be reasonably successful, the only possible approach is a multidisciplinary one. Many of the rural water supply projects have failed because of technical and organizational problems, lack of trained manpower and lack of mutual understanding on the part of villagers, engineers and planners. This means that not only most-appropriate technologies be applied, but they should also be linked to the other sectors of rural development, including the education and awareness of villagers on the significance and proper use of water supply, and sanitary installations, however, simple they may be.

There is a need for Information System if water supply sector is required to obtain strategic information for planning purpose and for management and operational control. In designing of information system it is necessary to identify the relevant information items required to be included in the system, the level at which it is to be used, the details and the time frequency[16].

A Management Information System (MIS) processes raw data and provides information to the competent authority for planning, coordination, control and monitoring (Figure 9.5). A computerised MIS is a more sophisticated method of providing useful information in different formats to different levels of administrators for assisting them in discharging their duties in a more efficient manner.

REFERENCES

1. PHED (1997) : *An Overview of Water Supply & Sanitation,* Public Health Engineering Department, Govt. of Rajasthan, p. 27.
2. PHED (1997) : *Ibid.* p. 31.
3. PHED (1998) : *Drinking Water : An Overview,* Public Health Engineering Department, Govt. of Rajasthan, p. 15.
4. PHED (1999) : *Annual Report,* Public Health Engineering Department, Govt. of Rajasthan.
5. Chouhan, T.S. (1996) : "Water Resources of India : Appraisal, Conservation and Management", *Water Resources and Their Utilisation,* Vigyan Prakashan, Jodhpur, p. 58.

6. Web site : *http://www.rajirrigation.com/bisalpur_status.htm,*

7. *Ibid.*

8. *Ibid.*

9. Web site : *http://www.investrajasthan.com/business/bis52828.htm*

10. White, G.W. (1977) : Pattern of Domestic Water Use in Low Income Communities, Water, Waste and Health in Hot Climate, in Ed. R. Feachem, John Wiley & Sons, p. 110.

11. White, G.W. (1977) : *ibid,* p. 111.

12. Watt, J.et al. (1953) : "Diarrhoeal Disease in Fresno County, California", *Am. Journ. of Pub. Health,* 43, pp. 728-741.

13. Moore et al. (1966) : "Diarrheal Disease Studies in Costa Rica : 1, Plan and Method of Investigation", *Am. Journ. of Pub. Health,* 56, pp. 442-451.

14. Wittman et al. (1967) : "An Evaluation of the Relationship Between Nutritional Status and Infections by Means of a Field Study", *South African-Medical Journal,* 41, pp. 664-682.

15. Marchall, C.L. (1972) : "Some Exercises in Social Ecology : Health, Disease and Modernization in the Ryukyu Island", *In The Careless Technology,* ed. M.T. Farvar and J.P. Milton, Now York, Natural History Press, pp. 5-18.

16. Ghosh, G et al (1995) : Water Supply in Rural India, Ashish Publishing House, New Delhi, p. 231.

❑❑❑

10

Planning and Management of Water Resource

Water is essential to life and civilisation is something of a dialogue between man and water. The earliest human beings settled on river banks, besides lakes and on the sea coast, and there too flourished agricultural and much of industrial activity. The proper management of water, for domestic and economic purposes, is a basic aspects of social management and governance.[1]

India is blessed with one of the largest water holder of all countries of world.[2] Groundwater resources are abundant as India has large number of perennial rivers and tributaries that cover most parts of our country. India has been largely dependent on rains, which has resulted in uncertainty of cultivation in a dominantly agricultural country. Thus, in face of drought and famines, the water supply systems were devised and practised from the earliest times.

If we look into our past and in the *Vedas* and the *Puranas* like *Yazurveda, Shushrut* and *Charak Samhitas*, the concept of embodiment of GOD in water is very well and vividly reflected. Quotation in this context is well suiting :

आपोवा इदं सर्व विश्वभूतान्याय:

प्रणवापो ऽपशवाय: अन्य माप:

अमृत माप: संराडापो ऽविराड़ोप:

स्मुराड़ापो ऽपश्चन्धापो: गंसत्याप:

ज्योतिंर्ग श्याप: ऽयजुर्गश्याप:

सत्य माप: सर्वदेवता ऽपोवा युर्भुव सुवराय:

Whole world depends on this indispensable water. All animate and inanimate objects around us require water for their existence. Water is the need of food and animals. Water is full of vital energy and supremacy. Water is in cosmic state, water is knowledge, Truth, Liberty, Source of Energy, and Symbol of Party. Water is the source of any form of life and energy. **Yajurved 4000 B.C.**

In *Devipurana*, the measurements and the main features of a dam are given; it is called as *palisandha* or *pala* or the common word *pali* of the modern Bundelkhand in Central India. During the Pre-Mauryan period in *Dhammapada* (a *pali* classic), *nethikas* are explained as builders of canals and aqueducts. In 300 B.C. Vishnugupta Kautalya mentioned that canals and dams were to be protected since cultivation could not depend entirely on rain water. *Amarkosha* (4th Century A.D.) describes lands dependant on rain water as *devamatrika* and those irrigated from canals and rivers as *nadimatrika*.

So much so that genesis and existence of the whole universe in our great Dharmagrarthas have been assigned to the conglomeration of five fundamental elements (PANCHA KHSITI TATVA). Due to these embodiments, the pious concept of water on the horizons of Indian cultural and religious heritage, infiltrated down to the generations. Emperors and Kings, paid basic attention to upkeep the sanctity of water, and appropriate environment necessary for optimum development of plants and animals. Devices and rules in this perspective were laid down.

The relics of past shows that in Rajasthan there were well planned systems for conservation of water resources and drainage. In the Middle era, the lakes, bunds, tanks, *talao* were built with different aims. But the principle aims included.

- independent, sufficient storage capacity, (to cater even for years together) replenishable from rains for drinking purposes.
- huge water storage in and around to protect the Kingship against the raids and attacks. In emergencies to kill and create a panic in the enemy forces by dint of sweep away

flow of water by opening the bottom valves of gigantic tanks located inside the forts.

- irrigation for cultivation of crop and food grains.
- recreational purposes such as duck shooting, hunting, bathing etc.
- religious purposes such as Ashwamegha Yajna, Holy celebrations etc.

Planning and management of water resource is of paramount significance. Water management is a practice which includes the integrated process of intake, conveyance, regulation, measurement, distribution, application and use of water to farms and domestic purposes with proper amount and at the right time, flood control, reclamation of waterlogged and flood affected areas, construction of check dams, etc.

10.1 WATER RESOURCE MANAGEMENT

Semi-arid areas are home to one sixth of the world's population.[3] Most of these are poor agro-pastoralists who depend totally on the renewable natural resources of water for their livelihoods. The inhabitants of this region are among the poorest people in the world. This poverty is partly caused by inadequate availability of water for crop, livestock and other enterprises. However, the shortage of water is not caused by low rainfall as normally perceived but, rather by a lack of capacity for sustainable management and use of the available rainwater.

The most critical management challenge is how to deal with the poor distribution of rainwater leading to short periods of too much water and flooding, and long periods of too little water. The question is, "can better management of the available rainwater help to reduce the occurrence and mitigate the impact of droughts during period or in places with low rainfall?"[4]

Precipitation (rainfall) is in general the most important source for the replenishment of the water resources in India.[5] In the season of rainfall it recharges the dams, reservoirs and other natural water resource, thus providing domestic water, water potential for irrigation and other purpose in Jaipur district. The main problem arises in the dry season, when intense heat and non-availability of rain water increases the high water requirement. The few small streams, leaving aside big stream, have some discharge at that time also, and if one can succeed in checking that small flow through small check dams constructed along the river courses at certain distances then they will serve as small reservoirs throughout the river course. By certain arrangement, the surplus water can thus be transferred to those areas where no discharge in streams is available or in areas having no canal. Such type of areas are maximum in Jaipur district.

10.2 ALLOCATION OF WATER RESOURCES

Allocation of water resource is done according to the basic needs of human beings. The most important use of water is that, It gives life to a thirsty man and then to meet his other requirements; allotment of reasonable amount of water for domestic purposes is the first consideration.

Table 10.1 : Annual Requirement of Water for Jaipur District

No.	Description	Water Demand (MLD)				
		2001	2011	2021	2031	2041
1	Domestic Use					
	a. Urban Water Supply	368.55	504.92	660.32	796.96	836.80
	b. Rural Water Supply	26.82	29.51	32.16	34.73	36.47
2	Industrial Use	73.71	100.98	132.06	159.39	167.36
3	Miscellaneous	117.27	158.85	206.14	247.77	260.16
4	Losses in Storage and Transmission	146.59	198.56	257.67	309.71	325.20
	Total Water Demand	732.95	992.82	1288.34	1548.56	1625.99

Source : Public Health Engineering Department, Jaipur.

At present, annually about 395.37 MLD (26.82 MLD for Rural Water Supply and 368.55 MLD for Urban Water Supply) (Table 10.1) of water is being utilised for domestic purposes in Jaipur district. This amount is rather too low for providing a reasonable standard of living to the district's population. The water need for domestic purpose in 2011 is expected to increase to 534.43 MLD (29.51 MLD for Rural Water Supply and 504.92 MLD for Urban Water Supply) (Figure 10.1).

Least attention was paid to water for industrial use, but some industries growing in rural areas certainly needs water. The most important industry developing in the area is carpet making which

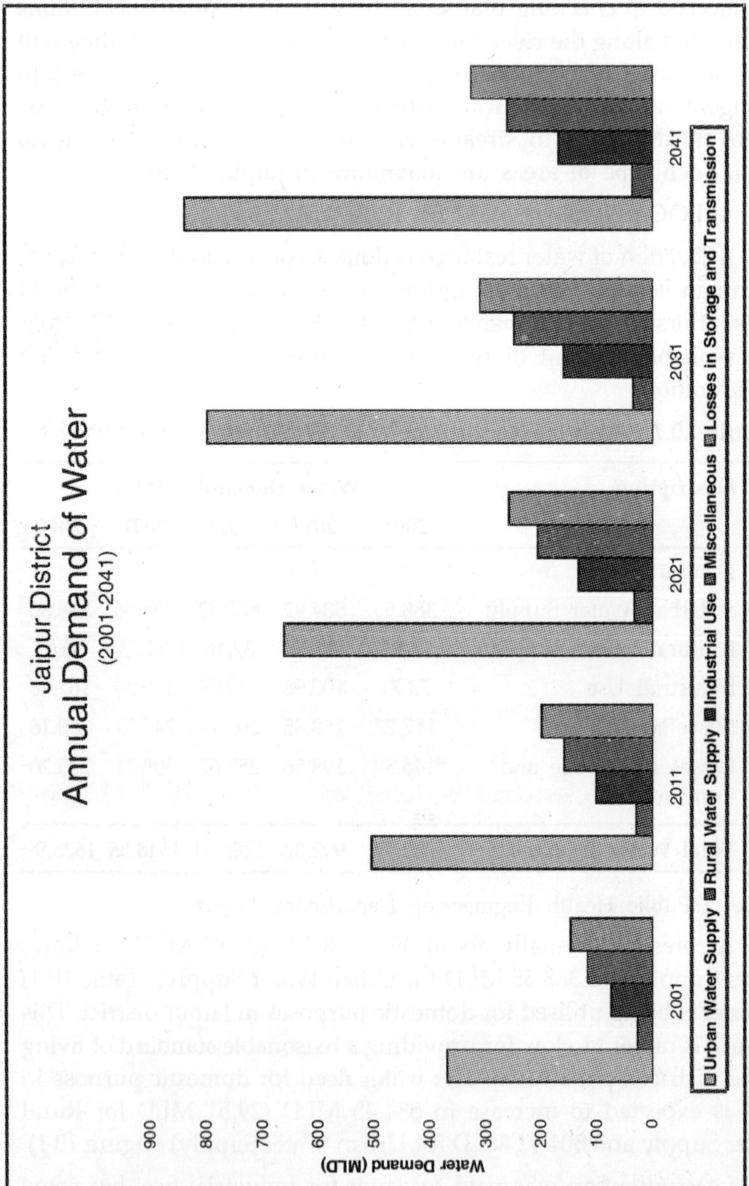
Jaipur District
Annual Demand of Water
(2001-2041)
Water Demand (MLD)
900
800
700
600
500
400
300
200
100
0
2001
2011
2021
2031
2041
Urban Water Supply
Rural Water Supply
Industrial Use
Miscellaneous
Losses in Storage and Transmission

Figure : 10.1

requires a large amount of water for colouring and washing. At present, it accounts for about 73.71 MLD annually. With the rapid development of industries, particularly paper, dye-based and cottage industries, the water allocation for 2011 will be about 100.98 MLD.

The total demand of water for domestic sector will be about 395.37 MLD. To fulfil the demand of domestic sector, as stated earlier, the government has decided to provide safe and adequate water through pipe line in each and every village with priority of most source villages. One major drawback in this scheme is the lack of consideration for livestock, where in every country where similar schemes are in progress, consideration for livestock is taken essentially. For the success of such schemes, the arrangement of revolving fund programme is necessary, so at the crisis period it can be continued (Figure 10.2).

10.3 RECYCLING AND REUSE OF WATER TO MEET INCREASING DEMAND OF WATER

Trying to work assiduously to make socio-economic progress with a view to find and restore a good quality of environment for teaming millions of our people in Rajasthan, we are beset with the problem of drought year after year.[6] Intriguing as it is for the Government to find employment for large number of people in the rural area, it calls for sweeping changes in our approach to tackle the situation. So far in implementing schemes to bring water to Rajasthan from Punjab through various Canal systems including the Indira Canal system, the Government has acquitted well by giving so much needed element for desert in Rajasthan. Welcome as all this is, it is neither adequate nor without problems of finding large financial resource. The ecology in sub- arid regions is fragile and for regeneration a continuous supply of water is needed to retain the moisture in the soil in certain low lying areas. One approach which is sometimes recommended is to construct small bunds, anicuts and reservoirs to artificially store water during rainy season and to release it gradually during dry periods for utilisation by more sophisticated methods of irrigation involving drip irrigation and retention of soil moisture over longer periods by providing a impervious layer of PVC lining a few inches below the ground level. At times bending the top soil with clay may also be recommended. Given the multiple use of water, it is needed not only for raising of crops but also for meadows, animal husbandry and for industry. In view of non-availability of

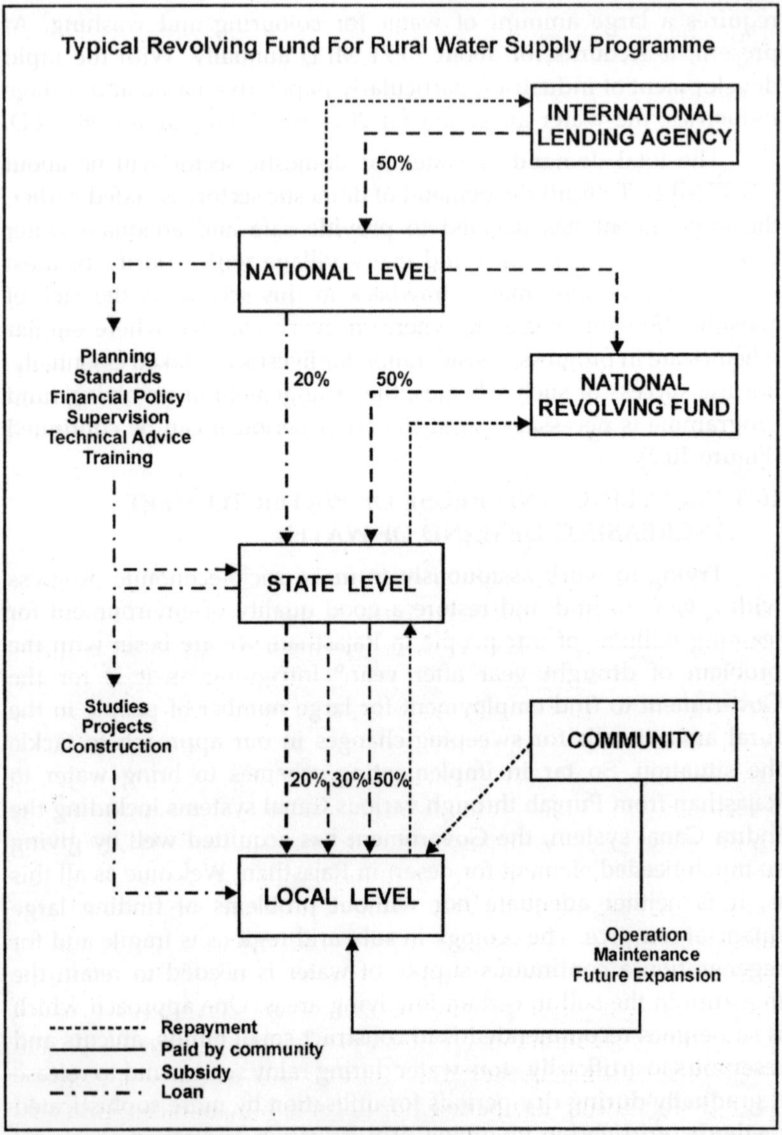

Figure : 10.2

fresh water for all these requirements it reflects on one's ability to use it over again by resorting to recycling and reuse as far as possible. It is necessary to stress that the problem can not be solved by means of grandiose schemes for National Water Grids and Garland Canals. More modest measures such as local watershed conservation and development and construction of bunds, check dams and percolation tanks all designed to store water where it falls short of - are far more relevant. It is also desirable to enmesh and infuse relief with plans aimed at drought and flood prevention. Involvement of the community into these programmes by bringing in the concept of equity in the sharing of available National resources also is essential.

The topography of Jaipur is such that a ridge line running from Surajpole to Chandpole, divides the walled city into two drainage zones. The area north of the ridge line drains towards an artificial lake Jalmahal, formed due to the construction of Mansagar dam. A major portion of north zone has a sewerage system, and the waste water from about 2.5 lac population is conveyed through a 1100 mm diameter outfall sewer, into a 27 MLD capacity sewage treatment plant, based on an "Extended Aeration System". The effluent obtained from this plant, flows into the lake Jalmahal (Man-sagar dam). An old canal system, downstream of the dam, irrigates a large area upto a distance of 12 kms. The sub-soil strata downstream of the dam, for a distance of about 3 Kms. consists of boulder formations upto a depth of about 15 meters, and the small basin surrounded by hills on three sides, is ideal for tapping underground water of high yield.

The catchment of Ramgarh lake, has received poor rainfall for the past four monsoons. The first poor rainfall in 1999, had resulted in a crisis of water supply for Jaipur. It was in the 1999, when during a search for alternative sources, the underground basin downstream of Mansagar was tapped by constructing an open well, about 1 Km. away from the toe of the dam. This well, which had a water column of 12 meters, yielded about 4.5 MLD of water. The good quality of water and the high yield obtained from this well led to construction of 2 more wells in 2000.

The Jalmahal lake receives an average of about 11 MLD of treated sewage from the treatment plant. Practically 75% of this quantity finds its way into the unlined canal system which apart from irrigation, also recharges the underground water table. Four wells have been constructed alongside this canal at suitable distances and water is being drawn from these wells at rate of an average of 10 MLD for the last about 2 years. It is thus established that major quantity of the treated waste from the north zone is being recycled. In terms of the total supply about 7 percent quantity is being recycled. The water obtained is potable having chemical parameters as shown in table 10.2.

Table 10.2 : Chemical Parameters of Recycled Jalmahal lake

No.	Chemical	Quantity
1	pH	8.1
2	Chloride	270 PPM
3	Nitrates	12 PPM
4	Total dissolved solids	1050 PPM

Source : Public Health Engineering Department, Jaipur.

However, since the water drawn from these open wells is fed through the common conveyance system from Ramgarh lake, it gets mixed with the lake water and is passed through the filtration process, before it is fed to the distribution system. It is thus established that nearly 10 million litters of waste water is being recycled and used for drinking water supply purposes in Jaipur.

On a wild guess, one is likely to be led to an erroneous conclusion about the unlimited supply of water available on our planet. Were it possible to spread the water available in the Oceans uniformally over the land surface, it would be possible to obtain approximately 2 Kms. depth of layer of water for use of mankind and other forms of life available on the earth? Looking to the present consumption of water, this quantity of water appears to be sufficiently large to last for many more years to come. The question arises as to whether all the water that is available in Sea and Oceans is useable for human consumption or not? What is the quantity of water that is actually available for use? Considering the annual average rainfall on our planet, the water contained in the Ice berg and thick layer of snow in the arctic and Antarctic regions, fresh water reserves on the ground, the underground water sources on land, the total quantity of fresh water is hardly 1 to 2 percent of the total water available. The replenishment that is taking place every year is even smaller and is hardly about 2% of the total quantity of water. This small quantity of water is not only limited but is fixed and cannot serve our requirement because fresh requirement of water is increasing every year on account of increase in population, increases in per capita consumption and increased activities of modern irrigation and industrialisation taking place at a very fast rate. One has to think and one suitable method that seems possible is to reuse the water repetitively for different purpose and consider recycle for similar use as far as possible. Besides conserving the quality and quantity of the fresh water, it should be possible to separate the material that is added after every use and realise the material so added as well as to recycle the water. This can enable not only the reuse of water but the recovery of raw-material added into the water may also result into certain economic benefits that the proposition can become economically viable.

10.4 NATIONAL WATER POLICY

Water is a prime natural resource, a basic human need and a precious national asset.[7] Over a time, demands on water had led to the evolution of superior technology which require less provisions of water. Today, a water supply system comprises a wide variety of components viz., dams, channels, treatment plants, storage, reservoirs and complete distribution systems. Several governmental organisations both in the Central and State are involved in policy, planning, implementation and operation and maintenance of such systems.

10.4.1 Safety of Structures

There should be proper organisational arrangements at the national and State levels for ensuring the safety of storage dams and other water-related structures. The central guidelines on the subject should be kept under constant review and periodically updated and reformulated. There should be a system of continuous surveillance and regular visits by experts.

10.4.2 Ground Water Development

- There should be a periodical reassessment on a scientific basis of the ground water potential, taking into consideration the quality of the water available and economic viability.
- Exploitation of ground water resources should not exceed the recharging possibilities. Ground water recharge projects should be developed and implemented for augmenting the available supplies.
- Integrated and coordinated development of surface water and ground water and their conjunctive use, should be

envisaged right from the project planning stage and should form an essential part of the project.

- Over exploitation of ground water should be avoided near the coast, to prevent ingress of sea water into sweet water aquifers.

10.4.3 Water Allocation Priorities

In the planning and operation of systems, water allocation priorities should be broadly as follows :

- Drinking water
- Irrigation
- Hydro-Power
- Navigation
- Industrial and other uses.

However, these priorities might be modified if necessary in particular regions with reference to area under specific considerations.

10.4.4 Drinking Water

Adequate drinking water facilities should be provided to the entire population both in urban and in rural areas. Irrigation and multipurpose projects should invariably include a drinking water component. Wherever there is no alternative source of drinking water, drinking water needs of human beings and animals should be the first charge on any available water.

10.4.5 Irrigation

- Irrigation planning either in an individual project or in a basin as a whole should take into account the feasibility of land cost-effective irrigation options possible from all available sources of water and appropriate irrigation techniques. The irrigation intensity should be such as to extend the benefits of irrigation to as large a number of farm families as possible, keeping in view the need to maximize production.
- There should be a close integration of water use and land use policies.
- Water allocation in an irrigation system should be done with due regard to equity and social justice. Disparities in the availability of water between head-reach and tail-end farms

and between large and small farms should be obviated by adoption of a rotational water distribution system and supply of water on a volumetric basis subject to certain ceilings.

- Concerted efforts should be made to ensure that the irrigation potential created is fully utilised and the gap between the potential created and its utilization is removed. For this purpose, the command area development approach should be adopted in all irrigation projects.

10.4.6 Water Rates

Water rates should be such as to convey the scarcity value or the resource to the users and to foster the motivation for economy in water-use. They should be adequate to cover the annual maintenance and operation charges and a part of the fixed costs. Efforts should be made to reach this ideal over a period, while ensuring the assured and timely supplies of irrigation water. The water rates for surface water and ground water should be rationalised with due regard to the interests of small and marginal farmers.

10.4.7 Participation of Farmers and Voluntary Agencies

Efforts should be made to involve farmers progressively in various aspect of management of irrigation systems, particularly in water distribution and collection of water rates. Assistance of voluntary agencies should be enlisted in educating the farmers in efficient water use and water management

10.4.8 Water Quality

Both surface water and ground water should be regularly monitored for quality. A phased programme should be undertaken for improvements in water quality

10.4.9 Water Zoning

Economic development and activities including agricultural, industrial and urban development, should be planned with due regard to the constraints imposed by the configuration of water availability. There should be a water zoning of the country and the economic activities should be guided and regulated in accordance with such zoning.

10.4.10 Conservation of Water

The efficiency of utilisation in all the diverse uses of water should be improved and an awareness of water as a scarce resource should be fostered. Conservation consciousness should be promoted through education regulation, incentives and disincentives.

10.4.11 Flood Control and Management

There should be a master plan for flood control and management for each flood prone basin. Sound watershed management through extensive soil conservation, catchment area treatment, preservation of check-dams should be promoted to reduce the intensity of floods. Adequate flood-cushion should be provided in water storage projects wherever feasible to facilitate better flood management An extensive network for flood forecasting should be established for timely warning to the settlements in the flood plains, alongwith the regulation of settlements and economic activity in the flood plain zones, to minimise the loss of life and property on account of floods while physical flood protection works like embankments and dykes will continue to be necessary. The emphasis should be on non-structural measures for the minimization of losses, such as flood forecasting and warning and flood plain zoning, so as to reduce the recurring expenditure on flood relief.

10.4.12 Land Erosion by Sea or River

The erosion of land, whether by the sea in coastal areas or by river waters inland, should be minimised by suitable cost-effective measures. The States and Union territories should also undertake all requisite steps to ensure that indiscriminate occupation and exploitation of coastal strips are discouraged and that the location of economic activities in area adjacent to the sea is regulated.

10.4.13 Drought Management

Drought prone areas should be made less vulnerable to drought associated problems through soil moisture conservation measures, water harvesting practices, the minimisation of evaporation losses, the development of the ground water potential and the transfer of surface water from surplus areas where feasible and appropriate. Pastures, forestry, or other modes of development which are relatively less water-demanding should be encouraged. In planning water resource development projects, the needs of drought-prone area should be given priority. Relief works undertaken for providing employment to drought-stricken populations should preferably be for drought proofing.

10.4.14 Science and Technology

For effective and economical management of our water resources, the frontiers of knowledge need to be pushed forward in several directions by intensifying research efforts in various area, including the following:

- Hydrometeorology
- Assessment of water resources;
- Snow and lake hydrology;
- Ground water hydrology and recharge;
- Prevention of salinity ingress;
- Water-harvesting;
- Evaporation and seepage losses;
- Economical designs for water resource projects;
- Crops and cropping systems;
- Sedimentation of reservoirs;
- The safety and longevity of water related structures;
- River-morphology and hydraulics;
- Soil and materials research;
- Better water management practices and improvements in operational technology
- Recycling and re-use;
- Use of sea water resources;

10.4.15 Training

A perspective plan for standardised training should be an integral part of water resources development. It should cover training in information systems, sectoral planning, projects planning and formulation, project management, operation of projects and their physical structures and systems and the management of the water distribution systems. The training should extend to all the categories of personnels involved in these activities and also the farmers.

In view of the vital importance of water for human and animal life, for maintaining ecological balance and for economic and development activities of all kinds, and considering its increasing scarcity, the planning and management of this resource and its optimal, economical and equitable use has become a matter of the utmost urgency The success of the national water policy will depend entirely on the development and maintenance of a national consensus and commitments to its underlying principles and objectives.

Water resource development, by raising economic and social standards, will lead to reduction in diseases and also in improvement in health if suitable precautions are taken throughout project development and its implementation. On the basis of many impact studies done throughout the world it in postulated that the development of water-resource projects ultimately enhance and improve the well-being of man if the project is will planned. Taking this promise and analogy, the WHO is operating its programme of water resource development.

REFERENCES

1. Ghosh, G et.al (1995) : *Water Supply in Rural India,* Ashish Pulishing House, New Delhi, p. 1.
2. Reoprt (1959) : *India's Food Crisis and Steps to Meet it,* Ford Foundation Team Report, p. 142.
3. Website : *http://www.cgiar.org/iwmi/home/rainwater.htm,* CGIAR online, Consultative Group on International Agricultural Research.
4. Website : *http://www.cgiar.org/iwmi/,* CGIAR online, Consultative Group on International Agricultural Research.
5. Ibid.
6. PHED (1988) : *Status of Drinking Water Supply in Rajasthan,* PHED, Govt. of Rajasthan, Jaipur, p. 22.
7. Ghosh, G et.al (1995) : *Water Supply in Rural India,* Ashish Pulishing House, New Delhi, p. 48.

11

Summary and Conclusion

Jaipur district, the largest of its kind in Rajasthan State, is by far one of the oldest existing districts and was carved out by portion of the erstwhile Jaipur and Kishangarh States. Jaipur district is situated between 26° 25′ to 27° 51′ north latitude and 74° 55′ to 76° 10′ east longitudes. It occupies a coveted position and engulfs an area of 10876.78 sq. km. which is an estimated 3.17 percent to the total area of Rajasthan State. The Jaipur district has on its northern fringes the Narnol district of Haryana State, on its north-eastern boundaries lies Sikar district of Rajasthan, on its south eastern boundaries lie Dausa district of Rajasthan, in the south is Tonk district, on its south western boundaries lies Ajmer district and in the west lies Nagaur district.

The geological formation of the region exhibits mostly of quartzite's and dolomitic marble of the Precambrian era. The region is broadly a part of Aravalli origanic system which has existed as stable mass in the deep history of the earth. The rock types exposed in the district are Pre-Delhi granites, granitic, gneisies, schists and Alwar, Ajabgarh and Railo formations to the Delhi group. The major part of the district is covered by quarternary sediments. The district is not much rich in mineral resources.

On the basis of landsat mosaic, hill systems of the district have been classified into (i) Puranghat-Nahargarh range, (ii) Bishangarh-Baneti range and (iii) Banera-Talu range. All these hill ranges are found in the north and eastern parts of the district. Southern, eastern and western parts of the district are peneplains with sand cover upto a thickness of 3 to 6 Metres. This district has one fourth area under mini hill tracts (having 30 percent or more than 30 percent slope), small hill enclave and other hill area.

The soil in the district is by and large sandy but there are certain areas towards the east and southern parts of the district where the soil in most of the parts is either brown or a rich alluvial soil. On the basis of colours and texture the soils of Jaipur district have been classified into (i) Chomu Aeolion Association, (ii) Bassi Rajori association (iii) Naraina Dudu Association, (iv) Rajori Chaksu Association (v) Chandrana Padasoli Association (vi) Chandrana Bandikui Association and Padasoli Phagi association. Soils vary from sandy loam to clay loam and southern part of the district has deep calcareous sub-soils. The soils of the district are very low in nitrogen, medium in potash and rich in phosphorus.

The district has semi-arid climatic conditions. The district has a hot dry summer and a bracing cold season. The cold season is from December to February and is followed by the hot season from March to last week of June. The southwest monsoon which follows lasts till about mid September the period from mid September to about the end of November constitutes post-monsoon period. Occasional hailstorms and cyclonic rains are common during the winter months.

The normal annual rainfall in the district is 54.82 cm. The rainfall generally increases from the western part to the eastern part. But in the Amber region and Viratnagar region the rainfall is a little higher than the surrounding parts of the district due to forested hills. The variation of the annual rainfall from year to year is very large. Monthly rainfall intensity is high in the monsoon month i.e., in June 5.13 cm., July 18.27 cm., August 18.07 cm. and September 8.52 cm. and decreases towards the both pre and post monsoon months.

The average population growth rate was 35.10 percent in the district. The growth rate was the lowest in the tehsils of western and southern parts of the district viz. Sanganer, Phulera, Dudu and Phagi tehsils. These tehsils are agriculturally backward due to calcareous and saline soils. The pattern of rural population growth rate is almost the same as the aggregate growth rate. The maximum rural population growth rate was in Bassi (+43.96) followed by Shahpura tehsil (+40.22). While in Sanganer (-14.69) and Jaipur tehsil (-13.69), rural population has decreased rapidly due to expansion of urban area.

The urban population growth rate is double (49.41 percent) than that of rural population (25.80 percent) in the Jaipur district. Females are less than males throughout the district. In Jaipur district females have remained less in comparison to males (897 females per 1000 males). The same pattern is found in rural areas (915 females per 1000 males) as well as in urban areas (878 females per 1000 males). Maximum male-female ratio in rural area is found in the western part of the district (Phulera 930, Chomu 921 and Phagi 920 females per 1000 males) while very low ratio is in Shahpura (901 females per 1000) tehsils. Besides, Jaipur is the centre of higher education, industrials and many administrative offices of the State and Central Government are also situated. In tehsils of north-eastern part of the district male and female ratio is either medium or low.

The major portion of the district is characterised by undulating plains. They are characterised by monotonous landscape with a fairly thick alluvial cover. The different hydrographic units of alluvial terrain comprises of older and younger alluvial plains of Banganga, Morel, Sabi rivers, abandoned channel courses, blown sand etc.

The Jaipur district is drained by five river basins as Mendha, Sabi, Banganga, Mashi, Morel basin. Mashi basin and Morel basin are sub-basins of river Banas, which passes through districts of Tonk and Sawai Madhopur. Mashi river joins river Banas near Tonk and Morel river joins river Banas near Olwara in Sawai Madhopur district. Mendha river terminates in to Sambhar lake near Phulera in Jaipur district where salt manufacturing unit is installed. Sabi and Banganga basins cover the Northern and North Eastern parts of Jaipur district.

The total available run off from these five river basins has been estimated as 1549.20 Mcum; out of this a quantity of 1227.04 Mcum is utilised for irrigation. Therefore, only 336.70 Mcum of surplus water is available for further utilisation.

In case of Sabi river basin, the available surplus water is not in significant quantity, so that it can be utilised for drinking purposes. It is always necessary to maintain minimum flows in the river, so that the ground water gets recharged. Hence it is concluded that the tapping of Sabi river basin is of no use.

Mashi and Morel river sub-basins enough surplus water is available to utilise it further for drinking purposes. Some proposals have been worked out to store surplus water available from Mashi river sub-basin in Isarda Dam project. In case of Morel river sub-basin a large quantity of surplus water i.e. 336.70 Mcum is going waste and some proposals has been made by State Irrigation Department to utilise this water in proposed Hydel project at Chambal. But in view of the future population growth rate and future water demand for drinking purposes, it is recommended that the surplus water available from Morel river basin should be utilised to meet the drinking water needs of Jaipur district by constructing a dam across the river before it joins Banas river.

Jaipur district once again is faced by severe drought this year (2002). The droughts have been the recurring problem of this country and will continue to be so for many years to come. Information gathered from different sources of government (Agriculture, Relief, Ground Water and Irrigation department) suggest *that this year's drought is much more severe than the last two years.*

Actual annual rainfall has been very low in district for the last three years and has continuously decreased since 1999. During last three years (1999 to 2001) the actual annual rainfall is less than the normal annual rainfall. The position of water available in reservoirs clearly shows that there was no good/heavy down pour which could have resulted in surface flow to feed the reservoir. Last year (2000) by the end of 31st December only 4 tanks had over flown, but this year (2001) not even one tank has over flown out of the 111 tanks in the Jaipur district.

Ground water is the main source of irrigation and drinking water supply in the Jaipur district. Ground water differs from sub-surface water such as capillary water and soil water. The average depth of water table in the district lies within 30 m below ground level. Fluctuations in water table occurs due to variation in ground water recharge, subsurface drainage and groundwater draft from wells.

The rural areas particularly suffer from the problem of coverage as well as from the problem of excess chemical quality parameters in ground water sources. Jaipur district is identified as having one of the worst affected district in respect of endemicity of fluorosis disease in Rajasthan. The major source of water in the rural villages of Jaipur is the ground water. It contains a high concentration of fluorides. Though persistent efforts for improving the quality of ground water by way of resorting to defluoridation techniques have been made for various water supply schemes but the problem is still persisting, due to a increased use of ground water, which can be attributed to the rise in population.

In addition, recharge of ground water vis-a-vis the total usage is minimal owing to the scanty rainfall. In order to meet the requirement of water for various competitive usages the withdrawal from the ground water source is on an increase year by year whereby the availability of water is restricted to the rainfall. Jaipur district is frequented by droughts which further deteriorates the availability of the ground water. In comparison to the limited availability of ground water, the additional needs of the district are quite large and, therefore, priorities have to be allocated to various competitive needs. The other requirements of water viz. for irrigation, mining, power generation become secondary. Highest priorities of water, therefore, has to be allocated for drinking purpose. For any economic activity the input of human resources is primary and, therefore, for the healthy growth of people of the district, safe and wholesome drinking water is required to be supplied.

There are 2131 populated villages out of 2187 villages in Jaipur district; and 2126 villages have been covered with drinking water supply with PHED (Public Health Engineering Department) in the year 2000-2001 Five main types (Piped scheme, P&T scheme, Regional scheme, TSS and Handpump scheme) of schemes are being executed by PHED for water supply in rural areas.

In Jaipur district, only 2.16 percent villages use traditional source scheme, 5.73 percent villages use piped scheme, 6.19 percent villages use pump & tank scheme, 13.42 percent villages use regional scheme and 72.22 percent use handpump scheme. During the period of 1991 to 2001 the number of villages under Handpump scheme decreased from 1699 to 1539 villages. The maximum number of villages under this scheme decreased in Chaksu tehsil (from 207 to 149 villages), where main problem is of high fluoride content in underground water besides having low potentiality of ground water. In case of handpumps it has not been possible to check the quality parameters of the water being supplied.

The total area under forest in Jaipur district was 7.58 percent in year 2000. It has slightly increased in last five years. Fallow land area is decreasing from year 1996 due to less amount of rainfall and less soil moisture of Jaipur district. The maximum 16.81 percent fallow land area was found in Sanganer and minimum 5.43 percent in Viratnagar tehsil.

In Jaipur district about 60.27 percent area is under NSA. Crops have furnished directly or indirectly a significant share in the income of the district. The main crop of the district is the rabi crop grown in the month of October and harvested in March. The crop consists of grains like wheat, barley and gram. Barely replaces wheat on the irrigated lands where the soil is light, irrigation water is scanty or its mineral content is high. When the monsoon are scanty then the area cultivated for barley increases. In the case of gram, generally irrigation is not required; it is also sown together with wheat and barley. The kharif crop also contributes to the supply of food grain to the district. It is generally sown in May-June end and harvested in September. The main crops that are sown during this season are jowar, bajra, groundnut and maize.

During five years (1996 to 2000) the irrigated area under wells has increased by 4.44 percent. It further shows that the importance of irrigation is increasing continuously. The main reason of increasing irrigation by wells is the erratic nature of rainfall. In the Jaipur district irrigated area by canals (0.55 percent) and tanks (0.02 percent) was very low. During last five years (1996 to 2000) the irrigated area under canals has decreased by 3.79 percent and area under tanks decreased by 0.64 percent (Table 7.3). It shows the impact of drought since 2000.

RESULTS OF HYPOTHESIS

High water potential region extends in central, northern and eastern parts of the district (Table 4.4). It has fertile alluvial soils and high ground water potentiality in the basins of the Mendha, Bandi, Banganga, Sota and Sabi rivers (Table 4.1). It has high cropping intensity, more population density and good literacy percentage (Table 3.5). Besides these the area under forest with tree cover is also high (Appendix-5) and maximum used of regional and piped water supply schemes in this region (Table 6.1). It can be said that the hypothesis "The level of water resource development of any area depends upon the livelihood pattern of the inhabitants of the area", is true of Jaipur district.

Four tehsils of Jaipur region, viz. Amber, Sanganer, Bassi and Jaipur fulfil all the requirements of vegetables of the capital city

Jaipur. Farmers of this region grow crops throughout the year and take 4-5 vegetable crops in a year. In Sanganer floriculture alongwith other food crops has been practiced for the last several years. This region is a major producer of wheat, barley, bajra, vegetables, fodder crops and horticulture. It can be concluded that this has high potentiality for the agricultural industries. Keeping in view infrastructure facilities and water resource development in the region it may be said that the hypothesis "Water resource development has taken place in these areas where infrastructure is developed", is true of Jaipur district.

The region is composed of western and southern tehsils of the district, viz. Phulera, Chaksu, Phagi, and Dudu has poor status in the water resources. It has low cropping intensity and region has problems of salinity and alkalinity. The soil of this area is calcareous, which is not suitable for agricultural practices. The quality and potentiality of ground water is also poor. This region having low population density and level of water resources development is also very low. Here it may be said that the hypothesis of the level of resources development is very low where the density of population is less is true in case of Jaipur district, while the, hypothesis "The level of resources development of any area depends upon the livelihood pattern of the inhabitants of the area" and "The present planning of resources development is politically oriented and it is not based on resources potentialities and necessities of inhabitants", are not true in case of Jaipur district.

SUGGESTIONS

- Suitable water harvesting procedures like run-off inducement, seepage and evaporation control measures in the area can conserve large quantities of water (Table 4.1). To minimise evaporation losses in the district, tree plantation near the ponds, reservoirs, and in the catchment areas should be done. The stored water can be used with advantage to give the crop life-saving irrigation. Moisture conservation in unirrigated areas is possible by increasing the depth of the soil moisture zones by suitable tillage practices.
- In order to develop the water resource of the State in the scientific manner, it is necessary to carry out water balance studies for the State as well as for the Jaipur district. At the

same time it is also necessary to determine problem areas with regard to the availability of safe and potable drinking water. The needs of drinking water for these problem areas will have to be met with by either creating of additional sources of water, or by using the limited resources by adopting efficient scientific principles and by recycling of water again after it gets polluted. After its first use, reusing the existing available resources of water and to work out possibility of diverting water from surplus areas to deficient areas can solve the problem to a great extent.

- The following Government Departments are responsible for co-ordination of activities with physical interventions that may be potentially responsible for the environmental conditions:
- Public Health Engineering Department
  - Irrigation Department
  - Soil Conservation Department
  - Industries Department
  - Forestry Department
  - Animal Husbandry Department
  - Ground Water Department
  - State Electricity Board Dept.
  - Agriculture Department
  - Mining Department

These Departments generally have irregular and inadequate communication with each other as a result of which the construction of dams done by Irrigation Department on the seasonal rivers adversely affects the planning of the other departments. The interception of water of these rivers, by diverting flow into the ground water, also adversely affects "the run off" for reaching in the existing reservoirs. Under the famine relief works lot of emphasis is laid for soil conservation works. Such works carried out in the catchment areas of the seasonal rivers divert the flow reaching the reservoirs constructed by the Irrigation department. As an illustration, the flow reaching Ramgarh Bund in Banganga basin are being quoted. This basin is already tapped by construction of Ramgarh Bund and under project action was taken for increasing the height of the dam to store more water. But the performance of this reservoir has not been satisfactory.

- The production from Ramgarh dam fell down considerably and only little additional quantity of water was obtained by construction of tubewells. The inflows during the period 1975 to 1981 had been very good. But thereafter steep decline has taken place. This aspect is required to be examined in order to understand the cause of reduction of flow in Ramgarh Bund so that mitigative measures can be undertaken. One reason may be that the catchment area is being used for cultivation of crops by constructing small anicuts across the river and its tributaries. By such activities the ground water table gets depleted and the flows generated during monsoon are diverted to the sub-soil. *It is, therefore, necessary to have proper communication between different Departments of the State Government and to see that the activities of one department do not interfere with the activities of the other department and cause an adverse impact on environment.* The catchment area of the seasonal rivers should be protected from soil erosion and afforestation should be done without allowing agriculture in such areas.
- Extensive environmental changes are caused by wind erosion and by the changes in ground water levels and water quality. The Forest Department has introduced mitigative measures in the form of forest plantations. The Ground Water Department is very much concerned about the situation of aquifers and has prepared proposals for charging the aquifers by rainwater collection systems. As already indicated efforts to build a network to hold surface flow in the river basin have been made in the past. It is necessary to see that such works are not carried out within the catchment area of the river basins where these rivers have been used to form reservoirs for supplying water for drinking purpose.
- The mitigative measures by Forest department should be undertaken in the catchment area of the river basin. Because, loss of forest area causes a number of difficulties; like frequent famine, less rainfall, less recharge of ground water aquifers, increased temperatures, soil erosion, pollution and many more human problems like fuel, timber, fodder,

medicines, water shortages, etc., all these problems can be solved by undertaking wide scale forestry work. The forestation efforts so far have not yet brought changes to the ground water situation.

- There are obvious major environmental impacts related to ground water pumping for irrigation and domestic purposes in the study area. According to the Ground Water Department data, water use for irrigation is over three times more than for the domestic use. The ground water table is lowering in several locations due to over pumping of the aquifers. The pumping of ground water with high TDS values for irrigation increases the salt concentration in the land and adversely affects vegetation. The evaporation is rapid and the salt concentrates. Visible white traces of salt can be seen in some of the ground water irrigated areas in Chaksu, Dudu, Phagi and Phulera tehsils. Usually, if the soil is irrigated with salty water for 2-5 years, then the cultivation becomes impossible and the soil is left for a few years. Depending on the soil type it is cleaned of salinity by rainwater in a few years and can again be cultivated. There is evident need to survey and introduce irrigation systems which do not consume much water and do not cause an increase in salinity in the soil. Local applications for drip irrigation should be tested and developed. Significant water conservation together with better fertility can be achieved through installation of sub-surface irrigation pipes. They will neither cause salt concentration nor will they consume large quantities of ground water.
- The area of fluoride and good water seems to be very narrow. There are abrupt changes in quality of water. An open well in the middle of village is polluted while a tube well little away from it is not polluted. In Sambhar the quality of water varies abruptly; in one well the water is saline and in the other which is at a short distance the water is potable. The quality of water in Shivdaspura is having excess fluorides but good quality water is available at a short distance. The parts of high fluoride concentrations on its border have good quality water. These fluoride affected aquifers may work as barriers for the flow of ground water and where such barriers are available across the sub surface flows, good quality of

water can be tapped on its upstream site. The fluoride concentrations are evidently following the ground water flow directions. Quality aquifers can potentially be found even few hundreds metres away from the fluoride wells. The direction of the ground water flow needs to be known when planning the well fields. There is no evidence of changing of fluoride and nitrate values as a function of depth. A scientific survey for defining the origin of the polluting components and their connections to the ground water as well as the treatment possibilities and needs should be carried out in co-operation with relevant National and International institutions.

- While there is an increasing demand for ground water in rural and urban water supply, most of the aquifers are already over pumped (Table 4.4 and Appendix-10). The over pumping is mainly due to a demand for ground water for irrigation. New boreholes are drilled by private drilling contractors without careful control, which should be checked from time to time by the State government.
- The ground water resources data show that there is limited scope for further development of groundwater for irrigation purposes (Table 4.4). All the hydrogeological formation of the entire district may have to be covered by micro level surveys to find out the middle and south western part of the district, which can be developed in future.
- High concentration of fluoride in groundwater is serious problem in Chaksu, Phagi and Phulera tehsils areas of the district. Detailed study has to be carried out to find out the sources of high concentration of fluoride in groundwater and also the change in concentration of fluoride in ground water with depth.
- In Jaipur district the availability of groundwater is a serious problem and day by day ground water level is going down. So the State Govt. plans have to be implemented properly to solve the groundwater problem to a great extent.
- Adequate power supply for energisation of pump-sets would reduce the financial burden of farmer utilising ground water resource and inturn boost the agricultural activity in the

district. Diesel operated pump sets unnecessarily enhance the cost of tubewell water.

- When introducing irrigation dams for intensive farming, care should-be taken not to harmfully restrict the surface and subsurface flow of water through seasonal streams. The management of surface water resources will also require active community participation.
- Water near the Sambhar Salt lake, Jobner, middle part of Phagi tehsil, eastern part of the Dudu, Chaksu tehsil, are not suitable for drinking. In these areas surface water should be harnessed and utilised.
- In case of rural area wherever quality parameters of ground water permit construction of handpumps, it is suggested that the coverage should be done by providing handpumps. It is only in respect of such villages (Table 5.4), which have quality problem that it is suggested that these should be covered by regional schemes. The regional schemes for these area will in fact be the extension schemes of the main village. Since the fluoride problem has been given highest priority in case of Jaipur district the regional water supply schemes for solving the problem should be executed on top priority.
- The major drawback in the present water supply schemes is that, there is no provision of water supply for livestock. Also, the release of funds in instalment for construction of the new schemes impedes rapid development. A scheme is started with a good sum of amount (say about 25 percent cost of the total estimate) and after utilization of first instalment, the contractor has to wait for another instalment of fund and if in the meantime, the cost of the materials has increased, then he is required to submit another estimate which consumes further more time, and creates hindrances in the execution of the schemes. For the success of such type of schemes, it is necessary to make an arrangement of Revolving Fund Programme, so that the crisis period can be overcome without such difficulty and scheme is completed within the time- schedule.

❑❑❑

Appendix - 1

Rainfall data recorded at all rain-gauge stations in year 1999 to 2001 along- with Normal and Percentage Deviation from normal of Jaipur District Actual Annual Rainfall and their difference of Jaipur district

			1999		2000		2001	
		Normal Annual Rainfall (cm.)	Actual Annual Rainfall (cm.)	Difference with normal in (+) and (-)	Actual Annual Rainfall (cm.)	Difference with normal in (+) and (-)	Actual Annual Rainfall (cm.)	Difference with normal in (+) and (-)
1	2	3	4	5	6	7	8	9
1	Jaipur	56.38	47.70	-8.68	34.62	-21.76	32.62	-23.76
2	Bassi	49.76	37.00	-12.76	31.20	-18.56	30.60	-19.16
3	Chaksu	58.39	46.00	-12.39	38.70	-19.69	31.66	-26.73
4	Sanganer	57.18	31.60	-25.58	45.50	-11.68	48.70	-8.48
5	Amer	61.80	41.00	-20.80	29.30	-32.50	35.90	-25.90
6	J. Ramgarh	54.73	47.20	-7.53	56.90	2.17	38.70	-16.03
7	Chomu	52.63	35.50	-17.13	34.10	-18.53	40.90	-11.73
8	Samod	52.27	40.20	-12.07				
9	Sambhar	47.87	58.60	10.73	59.02	11.15	38.47	-9.40

Contd...

...Contd.

1	2	3	4	5	6	7	8	9
10	Phulera	42.75	40.30	-2.45	22.68	-20.07	37.10	-5.65
11	Naraina	49.37	20.70	-28.67	19.16	-30.21	38.69	-10.68
12	Dudu	47.81	35.80	-12.01	23.78	-24.03	24.69	-23.12
13	Phagi	48.66	41.50	-7.16	28.06	-20.60	36.78	-11.88
14	Kotputli	51.58	47.90	-3.68	50.20	-1.38	42.58	-9.00
15	Pawata	52.41	57.00	4.59	74.90	22.49	48.80	-3.61
16	Shahpura	59.28	50.00	-9.28	40.80	-18.48	39.90	-19.38
17	Viratnagar	59.39	59.40	0.01	47.20	-12.19	49.35	-10.04

Source : Indian Meteorology Department, Jaipur.

Appendix - 2

Soil Test Classes and Nutrient Index Value

(Percentage of soil samples)

No.	Name of Panchayat Samiti	No. of soil samples Analysed	Ph		EC		OC				P_2O_5				K_20			
			N	A	N	S	L	M	H	NI	L	M	H	NI	L	M	H	NI
1	Bassi	1022	83.9	16.1	98.5	1.5	93.8	5.6	0.6	1.07	7.7	20.0	72.3	2.65	5.5	94.5	-	1.94
2	Sanganer	678	67.4	32.6	98.0	2.0	97.0	2.5	0.5	1.03	26.5	48.5	25.0	1.98	2.0	85.0	13.0	2.11
3	Jhotwara	2078	78.9	21.1	97.6	2.4	97.3	2.5	0.2	1.03	7.0	31.0	62.0	2.55	1.0	90.5	8.5	2.08
4	Phagi	604	69.1	30.9	98.4	1.6	97.5	2.5	-	1.02	15.5	33.6	50.9	2.35	-	79.0	21.0	2.21
5	Sambhar Lake	492	74.9	25.1	97.4	2.6	99.0	1.0	-	1.01	9.8	21.0	69.2	2.59	4.8	95.2	-	1.95
6	Dudu	309	59.4	40.6	95.8	4.2	98.0	2.0	-	1.02	16.4	28.0	55.0	2.39	2.5	91.0	6.5	2.04
7	Govindgarh	694	80.2	19.8	98.5	1.5	99.5	0.5	-	1.01	23.4	35.6	41.0	2.18	9.3	90.7	-	1.90
8	Amer	520	70.0	30.0	99.2	0.8	96.0	3.6	0.4	1.04	14.6	31.3	54.1	2.39	4.0	73.4	22.6	2.18
9	Jamwa Ramgarh	599	72.8	27.2	99.5	0.5	95.2	4.8	0.5	1.05	8.2	16.7	75.1	2.67	3.5	88.0	8.5	2.05
10	Shahpura	187	87.2	12.8	100.0	-	89.3	10.7	-	1.11	14.4	16.0	69.6	2.55	-	78.2	21.8	2.14
11	Viratnagar	452	79.2	20.8	98.2	1.8	98.8	1.0	0.2	1.01	5.1	15.1	79.8	2.74	4.0	75.0	21.0	2.17
12	Kotputli	821	83.0	17.0	98.8	1.2	97.3	2.7	-	1.03	5.1	27.0	67.9	2.63	1.5	90.0	8.5	2.07
13	Chaksu	518	73.9	26.1	95.9	4.1	97.1	2.9	-	1.03	18.6	33.5	47.9	2.29	3.5	67.5	29.0	2.25
	Total	8974	75.4	24.6	98.1	2.0	96.6	3.3	0.4	1.0	13.3	27.5	59.2	2.5	3.8	84.5	16.0	2.1

Limits for Nutrient Index (NI) Organic Carbon, available phosphorous and potash

Very Low (VL) : upto - 1.2, low : 1.2-1.67 medium : 2.01-2.33 Towards medium :1.68-2.00 High : above 2.33

Source : Department of Agriculture, Rajasthan.

Appendix - 3

Road Length in Rajasthan (2001-2002)

(in K.M.)

No.	District	National Highways	Painted	Metalled	Gravelled	Fair	Total
1	Ajmer	242	2519	83	57		2901
2	Alwar	35	3171	139	166	-	3511
3	Banswara	-	1486	128	133	-	1747
4	Barmer	189	3900	205	815	-	5109
5	Bharatpur	71	1874	36	4	-	1985
6	Bhilwara	129	2445	375	921	13	3883
7	Bikaner	325	2884	111	294	10	3624
8	Bundi	107	1149	87	183	-	1526
9	Chittorgarh	202	2264	178	187	-	2831
10	Churu	239	2490	165	116	-	3010
11	Dholpur	28	978	83	44	-	1133
12	Dungarpur	34	1224	121	72		1451
13	Ganganagar	124	2976	7	105	-	3212
14	Jaipur	401	3447	146	66	42	4102
15	Jaisalmer	196	2770	58	520	-	3544
16	Jalore	41	2121	186	294	18	2660
17	Jhalawar	121	942	135	162	-	1360
18	Jhunjhunu	-	2048	280	164		2492
19	Jodhpur	218	4348	182	1077	-	5825
20	Kota	120	1382	91	157	2	1752
21	Nagaur	330	3319	602	787		5038
22	Pali	218	2864	419	1043		4544
23	Sawai Madhopur	-	938	98	160		1196
24	Sikar	191	1835	163	327		2516
25	Sirohi	143	1119	77	101	-	1440
26	Tonk	109	1365	54	234	8	1770
27	Udaipur	250	2841	200	151	7	3449
28	Baran	143	967	64	306	11	1491
29	Dausa	88	1134	61	225		1508
30	Rajsamand	155	1263	185	155	-	1758
31	Hanumangarh	-	2547	4	68		2619
32	Karauli	4	1218	163	99	2	1486

Source : Directorate of Economic and Statistics, Jaipur.

Appendix - 4

Bureau of Indian Standard (BIS) for Drinking Water (IS 10500 – 1983)

S.No.	Substance or Characteristics	Requirement / desirable limit
1	Colour, Hazen unit, Max.	10
2	Odour	Unobjectionable
3	Taste	Agreeable
4	Turbidity, JTU, Max.	10
5	pH value	6.5 - 8.5
6	Total Hardness (as CaCO3) mg/L, Max.	300
7	Calcium (as Ca) mg/L, Max.	75
8	Magnesium (as Mg) mg/L, Max.	30
9	Copper (as Cu) mg/L, Max.	0.05
10	Iron (as Fe) mg/L, Max.	0.3
11	Manganese (as Mn) mg/L, Max.	0.1
12	Chlorides (as Cl) mg/L, Max.	250
13	Sulphate (as SO_4) mg/L, Max.	150
14	Nitrate (as NO_3) mg/L, Max.	45
15	Fluorides (as F) mg/L, Max.	0.6 - 1.2
16	Phenolics (as C_6H_5OH) mg/L, Max.	0.001
17	Mercury (as Hg) mg/L, Max.	0.001
18	Cadmium (as Cd) mg/L, Max.	0.01
19	Selenium (as Se) mg/L, Max.	0.01
20	Arsenic (as As) mg/L, Max.	0.05
21	Cyanide (as CN) mg/L, Max.	0.05
22	Lead (as Pd) mg/L, Max.	0.1
23	Zinc (as Zn) mg/L, Max.	5.0
24	Anionic Detergent (as MBAS) mg/L, Max.	0.2
25	Chromium (as Cr^{6+}) mg/L, Max.	0.05

Contd...

...Contd.

S.No.	Substance or Characteristics	Requirement / desirable limit
26	Poly-nuclear Aromatic Hydrocarbons (as PAH) mg/L, Max.	-
27	Mineral Oil, mg/L, Max.	0.01
28	Residual Free Chlorine, mg/L, Max.	0.2
29	Pesticides	Absent
30	Radioactive	-
	(a) Alpha emitters m Ci/ml, Max.	10^{-8}
	(b) Beta emitters m Ci/ml, Max.	10^{-7}

Indian Council of Medical Research (ICMR) standards for Physical & Chemical Characteristics of Drinking Water

S.No.	Substance or Characteristics	Highest Desirable Level	Max. Permissible Level
1	Colour, Hazen unit	5	25
2	Odour	Not Disagreeable	Not Disagreeable
3	Taste	Not Disagreeable	Not Disagreeable
4	Turbidity, JTU	5	25
5	pH value	7 - 8.5	6.5 - 9.2
6	Total Hardness (as CaCO3) mg/L	300	600
7	Calcium (as Ca) mg/L	75	200
8	Magnesium (as Mg) mg/L	Not More than 50 gm/L	-
9	Copper (as Cu) mg/L	0.05	1.5
10	Iron (as Fe) mg/L	0.3	1.0
11	Manganese (as Mn) mg/L	0.1	1.5
12	Chlorides (as Cl) mg/L	200	1000
13	Sulphate (as SO_4) mg/L	200	400

Contd...

...Contd.

S.No.	Substance or Characteristics	Highest Desirable Level	Max. Permissible Level
14	Nitrate (as NO_3) mg/L	20	50
15	Fluorides (as F) mg/L	1.0	1.5
16	Phenolics (as C_6H_5OH) mg/L	0.001	0.002
17	Mercury (as Hg) mg/L	-	0.001
18	Cadmium (as Cd) mg/L	-	0.01
19	Selenium (as Se) mg/L	-	0.01
20	Arsenic (as As) mg/L	-	0.05
21	Cyanide (as CN) mg/L	-	0.05
22	Lead (as Pd) mg/L	-	0.10
23	Radioactive materials	-	-
	(a) Alpha emitters pCi/ml	-	3
	(b) Beta emitters pCi/ml	-	30

World Health Organisation (WHO) Standards for Physical & Chemical Characteristics of Drinking Water

S.No.	Substance or Characteristics	Highest Desirable Level	Max. Permissible Level
1	Colour, Hazen unit	5	50
2	Odour	Unobjectionable	Unobjectionable
3	Taste	Unobjectionable	Unobjectionable
4	Turbidity, JTU	5	25
5	pH value	7 - 8.5	6.5 - 9.2
6	Total Hardness (as CaCO3) mg/L	100	500
7	Calcium (as Ca) mg/L	75	200
8	Magnesium (as Mg) mg/L	230 if SO_4 is 250 mg/L upto 150 mg/L if SO_4 is less than 250 mg/L	230 if SO_4 is 250 mg/L upto 150 mg/L if SO_4 is less than 250 mg/L

Contd...

...Contd.

S.No.	Substance or Characteristics	Highest Desirable Level	Max. Permissible Level
9	Copper (as Cu) mg/L	0.05	1.5
10	Iron (as Fe) mg/L	0.05	1.5
11	Manganese (as Mn) mg/L	0.1	1.0
12	Chlorides (as Cl) mg/L	200	600
13	Sulphate (as SO_4) mg/L	200	400
14	Nitrate (as NO_3) mg/L	-	45
15	Fluorides (as F) mg/L	0.6 - 0.9	0.8 - 1.7
16	Phenolics (as C_6H_5OH) mg/L	0.001	0.002
17	Mercury (as Hg) mg/L	-	0.01
18	Cadmium (as Cd) mg/L	-	0.01
19	Selenium (as Se) mg/L	-	0.1
20	Arsenic (as As) mg/L	-	0.05
21	Cyanide (as CN) mg/L	-	0.05
22	Lead (as Pd) mg/L	-	0.1
23	Radioactive materials	-	-
	(a) Alpha emitters pCi/ml	-	3
	(b) Beta emitters pCi/ml	-	30

Drinking water standards : Physical & Chemical Characteristics of Drinking water of Ministry of Work and Housing

S.No.	Substance or Characteristics	Acceptable	Cause of Rejection
1	Colour, Hazen unit	5.0	2.5
2	Odour & Taste	Unobjectionable	Unobjectionable
4	Turbidity, JTU	2.5	10
5	pH value	7.0 - 8.5	6.5 - 9.2
6	Total Hardness (as CaCO3) mg/L	200	600
7	Calcium (as Ca) mg/L	75	200

Contd...

...Contd.

S.No.	Substance or Characteristics	Acceptable	Cause of Rejection
8	Magnesium (as Mg) mg/L	30	150
9	Copper (as Cu) mg/L	0.05	1.5
10	Iron (as Fe) mg/L	0.1	1.0
11	Manganese (as Mn) mg/L	0.05	0.5
12	Chlorides (as Cl) mg/L	200	1000
13	Sulphate (as SO_4) mg/L	200	400
14	Nitrate (as NO_3) mg/L	45	45
15	Fluorides (as F) mg/L	1.0	1.5
16	Phenolics (as C_6H_5OH) mg/L	0.001	0.002
17	Mercury (as Hg) mg/L	0.001	0.001
18	Cadmium (as Cd) mg/L	0.01	0.01
19	Selenium (as Se) mg/L	0.01	0.01
20	Arsenic (as As) mg/L	0.05	0.05
21	Cyanide (as CN) mg/L	0.05	0.05
22	Lead (as Pd) mg/L	0.1	0.1
23	Radioactive materials		
	(a) Alpha emitters pCi/ml	3.0	3.0
	(b) Beta emitters pCi/ml	30.0	30.0

Appendix - 5

Land Utilisation in Jaipur District-2002

(Area in Hectares)

No.	Tehsils	Total Area	Forest	Not available for cultivation			Other uncultivated land (Land Excluding Fallow Lands)				Fallow Lands			Total area sown
				Area under non-agricultural uses	Barren and unculturable land	Total (4+5)	Permanent Pasture and other grazing land	Land miscellaneous tree crop & groves	Culturable Waste Land	Total (7-9)	Fallow Lands other than Current Fallow	Current Fallow	Total (11-12)	
	1	2	3	4	5	6	7	8	9	10	11	12	13	14
1	Jaipur	52716	1028	12964	2597	15561	1695	216	2423	4334	2690	1342	4032	27761
	Area in %	100.00	1.95	24.59	4.93	29.52	3.22	0.41	4.60	8.22	5.10	2.55	7.65	52.66
2	Bassi	65469	2120	4699	3822	8521	4831	161	1707	6699	3753	4635	8388	39741
	Area in %	100.00	3.24	7.18	5.84	13.02	7.38	0.25	2.61	10.23	5.73	7.08	12.81	60.70
3	Chaksu	81177	89	6738	590	7328	7804	13	1777	9594	4162	5759	9921	54245
	Area in %	100.00	0.11	8.30	0.73	9.03	9.61	0.02	2.19	11.82	5.13	7.09	12.22	66.82
4	Sanganer	70197	1302	6960	2651	9611	3605	158	2547	6310	6392	5406	11798	41176
	Area in %	100.00	1.85	9.91	3.78	13.69	5.14	0.23	3.63	8.99	9.11	7.70	16.81	58.66

Contd...

...Contd.

1	2	3	4	5	6	7	8	9	10	11	12	13	14
5 Amer	89165	11393	4559	6785	11344	5917	177	1811	7905	5458	3980	9438	49085
Area in %	100.00	12.78	5.11	7.61	12.72	6.64	0.20	2.03	8.87	6.12	4.46	10.58	55.05
6 J. Ramgarh	103370	24188	4928	12501	17429	7195	89	3165	10449	3338	4624	7962	43342
Area in %	100.00	23.40	4.77	12.09	16.86	6.96	0.09	3.06	10.11	3.23	4.47	7.70	41.93
7 Chomu	68361	1477	3995	2548	6543	3813	14	522	4349	3037	2012	5049	50943
Area in %	100.00	2.16	5.84	3.73	9.57	5.58	0.02	0.76	6.36	4.44	2.94	7.39	74.52
8 Phulera	147048	181	8259	1900	10159	9522	36	3223	12781	6152	9185	15337	108590
Area in %	100.00	0.12	5.62	1.29	6.91	6.48	0.02	2.19	8.69	4.18	6.25	10.43	73.85
9 Dudu	133856	750	9808	1258	11066	13132	26	5483	18641	5587	15661	21248	82151
Area in %	100.00	0.56	7.33	0.94	8.27	9.81	0.02	4.10	13.93	4.17	11.70	15.87	61.37
10 Phagi	111434	662	8837	514	9351	11604	1	3894	15499	4729	10577	15306	70616
Area in %	100.00	0.59	7.93	0.46	8.39	10.41	0.00	3.49	13.91	4.24	9.49	13.74	63.37
11 Kotputli	81434	10274	4917	6000	10917	2859	11	1423	4293	2747	2508	5255	50695
Area in %	100.00	12.62	6.04	7.37	13.41	3.51	0.01	1.75	5.27	3.37	3.08	6.45	62.25
12 Viratnagar	48775	18141	1479	3666	5145	2205	14	1310	3529	1305	1344	2649	19311
Area in %	100.00	37.19	3.03	7.52	10.55	4.52	0.03	2.69	7.24	2.68	2.76	5.43	39.59
13 Shahpura	52557	7728	2980	3494	6474	4246	37	818	5101	2011	2594	4605	28649
Area in %	100.00	14.70	5.67	6.65	12.32	8.08	0.07	1.56	9.71	3.83	4.94	8.76	54.51

Source : *District Collector Office, Jaipur*

Appendix - 6

Tehsil-wise Area Under Major Crops in Jaipur District (2000)

Area in Hectares (Percentage)

No	Tehsil	Bajra	Jowar	Wheat	Maize	Barley	Gram
1	Jaipur	11224	10	12228	139	2849	2190
		(32.53)	(0.03)	(35.44)	(0.40)	(8.26)	(6.35)
2	Bassi	11360	119	17206	426	2704	2626
		(25.66)	(0.27)	(38.86)	(0.96)	(6.11)	(5.93)
3	Chaksu	7922	2185	13047	404	2228	3791
		(14.36)	(3.96)	(23.66)	(0.73)	(4.04)	(6.87)
4	Sanganer	13396	107	15246	204	3169	2442
		(31.36)	(0.25)	(35.69)	(0.48)	(7.42)	(5.72)
5	Amer	22939	9	19310	419	5547	6818
		(35.25)	(0.01)	(29.67)	(0.64)	(8.52)	(10.48)
6	J. Ramgarh	13353	154	14792	2041	2703	6492
		(27.03)	(0.31)	(29.94)	(4.13)	(5.47)	(13.14)
7	Chomu	21848	-	18897	246	7302	5097
		(30.79)	-	(26.63)	(0.35)	(10.29)	(7.18)
8	Phulera	32718	207	27173	31	12709	19285
		(27.03)	(0.17)	(22.45)	(0.03)	(10.50)	(15.93)
9	Dudu	10127	7818	7212	2134	3401	6997
		(11.10)	(8.57)	(7.91)	(2.34)	(3.73)	(7.67)
10	Phagi	6668	5787	10586	1132	2429	5043
		(9.22)	(8.00)	(14.64)	(1.57)	(3.36)	(6.98)
11	Kotputli	25571	52	14955	56	2374	7021
		(35.79)	(0.07)	(20.93)	(0.08)	(3.32)	(9.83)
12	Viratnagar	9667	18	7500	281	1524	3906
		(36.71)	(0.07)	(28.48)	(1.07)	(5.79)	(14.83)
13	Shahpura	16440	3	10338	193	2374	4939
		(41.52)	(0.01)	(26.11)	(0.49)	(5.99)	(12.47)

Source : *District Collector Office, Jaipur.*

Appendix - 7

Year wise Cropping Intensity in Jaipur District (1995-2000)

Area in Hectares
(Percentage)

Years	Gross Cropped Area	Net Sown Area	Double Cropped Area	Cropping Intensity
1995-96	747573	656305	91268	113.91
	(100.00)	(87.79)	(12.21)	
1996-97	825219	685398	139821	120.40
	(100.00)	(83.06)	(16.94)	
1997-98	891707	699247	192460	127.52
	(100.00)	(78.42)	(21.58)	
1998-99	870303	695982	174321	125.05
	(100.00)	(79.97)	(20.03)	
1999-2000	783999	666305	117694	117.66
	(100.00)	(84.99)	(15.01)	

Source : *District Collector Office, Jaipur.*

Appendix - 8

Cropping Intensity in Jaipur District

Area in Hectares (Percentage)

No.	Tehsil	Gross Cropped Area	Net Sown Area	Double Cropped Area	Cropping Intensity
1	Jaipur	34499	27761	6738	124.27
		(100.00)	(80.47)	(19.53)	
2	Bassi	44277	39741	4536	111.41
		(100.00)	(89.76)	(10.24)	
3	Chaksu	55151	54245	906	101.67
		(100.00)	(98.36)	(1.64)	
4	Sanganer	42717	41176	1541	103.74
		(100.00)	(96.39)	(3.61)	
5	Amer	65075	49085	15990	132.58
		(100.00)	(75.43)	(24.57)	
6	J. Ramgarh	49406	43342	6064	113.99
		(100.00)	(87.73)	(12.27)	
7	Chomu	70951	50943	20008	139.28
		(100.00)	(71.80)	(28.20)	
8	Phulera	121045	108590	12455	111.47
		(100.00)	(89.71)	(10.29)	
9	Dudu	91198	82151	9047	111.01
		(100.00)	(90.08)	(9.92)	
10	Phagi	72301	70616	1685	102.39
		(100.00)	(97.67)	(2.33)	
11	Kotputli	71447	50695	20752	140.94
		(100.00)	(70.95)	(29.05)	
12	Viratnagar	26332	19311	7021	136.36
		(100.00)	(73.34)	(26.66)	
13	Shahpura	39600	28649	10951	138.22
		(100.00)	(72.35)	(27.65)	
Total		783999	666305	117694	117.66
		(100.00)	(84.99)	(15.01)	

Source : *District Collector Office, Jaipur.*

Appendix - 9

Tehsil-wise Irrigated Area by Different Sources in Jaipur District

Area in Hectares
(Percentage)

	Tehsil	Wells & Tubewells	Canals	Tanks	Other sources	Net Irrigated Area
1	Jaipur	23452	55	0	0	23507
		(99.77)	(0.23)	(0.00)	(0.00)	
2	Bassi	25496	-	-	-	25496
		(100.00)	(0.00)	(0.00)	(0.00)	
3	Chaksu	19893	-	-	-	19893
		(100.00)	(0.00)	(0.00)	(0.00)	
4	Sanganer	23258	1415	-	-	24673
		(94.26)	(5.74)	(0.00)	(0.00)	
5	Amer	36592	202	-	-	36794
		(99.45)	(0.55)	(0.00)	(0.00)	
6	J. Ramgarh	21785	-	5	-	21790
		(99.98)	(0.00)	(0.02)	(0.00)	
7	Chomu	42160	-	-	-	42160
		(100.00)	(0.00)	(0.00)	(0.00)	
8	Phulera	64186	-	-	-	64186
		(100.00)	(0.00)	(0.00)	(0.00)	
9	Dudu	15013	-	-	-	15013
		(100.00)	(0.00)	(0.00)	(0.00)	
10	Phagi	19995	-	60	-	20055
		(99.70)	(0.00)	(0.30)	(0.00)	
11	Kotputli	34303	281	-	-	34584
		(99.19)	(0.81)	(0.00)	(0.00)	
12	Viratnagar	12864	23	-	-	12887
		(99.82)	(0.18)	(0.00)	(0.00)	
13	Shahpura	18194	-	-	-	18194
		(100.00)	(0.00)	(0.00)	(0.00)	
	Total	357191	1976	65	-	359232
		(99.43)	(0.55)	(0.02)	(0.00)	

Source : *District Collector Office, Jaipur.*

Appendix - 10

Block-wise Ground Water Potential in Jaipur District (2000)

No	Block	Zones	Depth of water table (m.)			Avg. Yield	Area
			Min.	Max.	Avg.		sq.km.
1	Amer	I	9	31	20	9000 lph.	746.46
		II	12.7	31.4	22.05	40000 lpd.	104.02
2	Viratnagar	I	3.6	23.4	13.5	93400 lpd.	588.37
		.II	11	13	12	6000 lph.	75.59
3	Bassi	I	12	13	12.5	65000 lpd.	-
		II	14	27	20.5	10000 lph.	-
		III	23.2	27.9	25.55	52500 lpd.	-
4	Chaksu	I	7.8	22.8	15.3	15000 lph.	552.06
		II	13.15	17.5	15.325	52500 lpd.	177.29
5	Dudu	I	10.7	21	15.85	57500 lpd.	327.26
		II	8.9	28.9	18.9	38000 lpd.	407.92
		III	6.4	19.6	13	38000 lpd.	1055.67
6	Govindgarh	I	9.9	31	20.45	15000 lph.	638.08
7	J. Ramgarh	I	5.6	28.5	17.05	85000 lpd.	577.82
		II	13	14	13.5	45000 lpd.	96.85
		III	16	17	16.5	45000 lpd.	78.95
		IV	20.5	25	22.75	40000 lpd.	136.00
		V	13	14	13.5	40000 lpd.	72.25
8	Jaipur	I	10.55	18.9	14.725	96500 lpd.	-
9	Kotputli	I	6.1	14.2	10.15	25000 lph.	427.32
		II	13.1	27.75	20.425	40000 lpd.	130.54
10	Phagi	I	4.5	30	17.25	60000 lpd.	396.11
		II	15	16	15.5	43400 lpd.	880.44
11	Sambhar	I	11.5	13.5	12.5	25000 lph.	249.44
		II	6.7	21.9	14.3	25000 lph.	635.55
12	Sanganer	I	16.5	23.9	20.2	10000 lph.	613.89
13	Shahpura	I	17.5	29	23.25	18000 lph.	384.13

Source : *Central Ground Water Board, Jaipur.*

Bibliography

Abstract (2001) : *District Statistics Abstract,* Directorate of Economics & Statistics, Rajasthan, Jaipur. p. 3.

Ahmed, K.S. (1946) : "Environment and the Institution of Population in India", *Indiana Geographical Journal,* Vol. VI, p. 33.

Arceivala, S.J. (1981) : *Wastewater Treatment and Disposal,* Marcel Dekker Inc., New York.

Arora, S.P. (1969) : "Planning a Rural Water Supply", *The Journal of Inst. of Engg. (India),* PH2, Vol.49, No. 6.

Arya, R.P. *et. al.* (2000) : *Road Atlas, Rajasthan,* Indian Map Service Publication, Jodhpur. p. 3.

Balasubramaniam, T. (1983) : "Application of Remote Sensing to Ground Water Studies-A Case Study from North Arcot District of Tamil Nadu, Proc. Nat. Symp. on Remote Sensing in Development and Management of Water Resources.

Beasdale, A. *et. al.* (1963) : "Study and Assessment of Water Resource", *Conservation of Water Resources,* Inst. Civil Engg. Symposium. pp. 121-136.

Bhati, V.M. (1996) : *Water Resources and Their Utilization,* Vigyan Prakashan, Jodhpur, p. 1.

Bhindouriy, B.P.S. *et. al.* (1986) : *Rural Development Strategy and Perspective.* Annual Publications, Delhi, India, p. 53.

Bijlani, H.C. *et. al.* (1990) : *Water Supply and Sanitation in India,* New Delhi, Oxford & IBH.

Bilas, R. (1980) : "Ground Water Resources of Varanasi District, India: An Assessment of Conditions, Use and Quality", *National Geog. Journal India,* Vol.26, p.81-93.

Bode, D.C. *et. al.* (1978) : *Biological effects of surfactants,* III. Environs. Pollut. 17:175.

Bradley, D.J. (1974) : "Water Supply: The Consequence of Change" (Elliott, K. and as Knight, J. eds.), In: *Human Right in Health,* Amsterdam : ASP North-Holland, pp. 81-98.

Buck, J.L. (1937) : *Land Utilization in China, Vol. 3. London.* p. 63.

Burton, I. (1974) : "Domestic Water Supplies for Rural People in the Developing Countries: The Hope of Technology", In *Human Rights in Health,* Eilliott, K. and Knight, J. eds., Ciba Foundation Symposium, 23 (New Series) Amsterdam, ASP -North Holland, pp. 61-79.

Census of India (1991) : *Jaipur District Census Hand Book.*

Chandran, R. (1977) : "The World Running Out of Water ", *The Illustrated Weekly,* Vol.24, p. 15.

Chandrashekhar, C.S. *et. al.* (1972) : *Growth Force in Regional Strategy.* University of Maysor, p. 55.

Chaturvedi, R. S. et. al (1985) : "Integrating Remote Sensing Technology in Ground Water Exploration-" Typical Case Study from Bundelkhand Region in U.P. Proc. Nat. Symp. on Remote Sensing in Development and Management of Water Resources.

Chouhan, T.S. (1996) : "Water Resources of India : Appraisal, Conservation and Management", *Water Resources and Their Utilization,* Vigyan Prakashan, Jodhpur, p. 58.

CPHERI (1972) : *Disinfections of Well Water (Technical Digest no. 26,* Nagpur, Central Public Health Engg. Res. Institute: also Pickford, J. (1977): "Water Treatment in Developing Countries", In *Water and Health in Hot Climate,* Feachem, R. *et. al.* eds., John Wiley, London, p. 186.

Department of Irrigation (1996) : *Report of the Working Group on Irrigation for The Ninth Five Year Plan* (1997-2002). Department of Irrigation, Govt. of Rajasthan, Jaipur.

Dept. of Agriculture (1995) : *Report of the Committee for the Fourth Five Year Plan on Agriculture Production - Rajasthan State,* Part I. p. 10

Dewangen, K. N. *et. al.* (1983) : "Use of Landsat Imagery in Ground Water Investigation in Part of Noyil Basin, Tamil Nadu," Proc. Nat. Symp. on Remote Sensing in Development and Management of Water Resources.

Dhabriya, S.S. *et. al.* (1986) : *Existing Land Use System and Possible Strategy to Evolve an Optimum Land Use Plan for Jaipur District.* Social Policy Research Institute, Jaipur, p. 15.

Dhar, G.M. *et. al.* (1971) : "A Study of Drinking Water in Village Rahimabad of Lucknow District", *Indian Journal of Med. Res.,* Vol.59, pp. 1922-1931.

Dieterich, B.H. *et. al.* (1963) : *Urban Water Supply Condition and Needs in Seventy Five Developing Countries,* WHO, Public Health p. 23.

Dixey, F (1950) : *A Practical Handbook of Water Supply,* Thomas Murby & Co., London, p. 304.

Donald, A.D. (1982) : *Soil and Land Use Planning, London.* p.49.

Ehorlich, R. Paul (1970) : *Population Resources Environment-Issues in Human Ecology* . W.M. Freeman and Co., San Francisco, p.38

Feachem, R. (1975) : "The Rational Allocation of Water Resources for the Domestic Need of Rural Communities in Developing Countries" in *Proceedings of Second World Congress of Water Resource,* New Delhi, Vol. 2 pp. 539-547.

FFT Reoprt (1959) : *India's Food crisis and steps to Meet it,* Ford Foundation Team Report, p. 142.

Folwell, A.P. (1917) : *Water Supply Engineering,* 3rd ed. John Wiley, New York, p. 23-52.

Fuller, M.L. (1910) : *Underground Wastes for Farm,* U.S. Geol. Survey, W.S. paper, p. 255.

Gardner, J.S. (1971) : *Physical Geography.* Harper College Press, London, p. 154.

Garg, S.K. (1992) : *Water Supply Engineering,* Tata Microhill, New Delhi, India.

Ghosh Shanker (1976) : "The Planning of our Forest Resources", *Yojna,* Vol. XX/16, p. 10.

Ghosh, G *et. al.* (1995) : Water Supply in Rural India, Ashish Publishing House, New Delhi, p. 231.

Sharma, K. P. *et. al.* (1996) : *Environmental Guidelines and Standards in India,* Technoscience Publication, Jaipur. p. 16.

GOI (1982) : *Evaluation of Rural Water Supply schemes in India by NEERI,* Nagpur sponsored by CPHEEO, MUD, Govt. of India

GOI (2001) : *Monsoon 2001,*Irrigation Department, Govt. of Rajasthan, Jaipur. p. 6

GOR (2002) : *A Note on Drought Situation in The State and Its Management (Samvat 2059),* Relief Department, Govt. of Rajasthan, p. 4-5.

GOR (2002) : *Interim Memorandum on Scarcity (Samvat-2059),* Relief Department, Govt. of Rajasthan. p. 1

Goument, C.E.V. (1920) : "Water Supply", *Sanitary Engineering,* Pt.1 Govt. Press (U.P.)

Gupta, V.P. *et. al.* (1970) : "A Study of Domestic Water Consumption in Nagpur", *The Journal of Inst. of Engg. (India),* Vol.51, PH1, pp. 5-7.

Hem, J.D. (1970). Study and Interpretation of the Chemical Characteristics of Natural Water, 2nd ed. U.S. Geol. Survey Water Supply Paper-1473, p. 363.

Heron, A.M. (1923) *The Geology of western Jaipur* : Records Geological Survey of India, Vol. 54.

Hussain, S. R. (1965) : *Text Book of Water Supply and Sanitation Engineering,* Oxford & USA Pub. Co., New Delhi, p. 93.

Joshi, B.H. (1992) : *An Analytical Approach to problems of Indian Agriculture : A Theoretical and System Approach,* B.R. Publishing Corporation, Delhi.

Kalbermatten, J. M. *et. al.* (1982) : *Appropriate Sanitation Alternatives : A technical and Economical Appraisal* (World Bank Studies in Water Supply and Sanitation, No. 1) Hopkins University Press.

Kalwar, S.C. *et. al.* (1991) : *Remote Sensing in Resource Geography,* Pointer Publisher, Jaipur.

Kalwar, S.C. (2008) : Wastelands and Planning for Development, Concept Publishing Company, New Delhi.

Kshirsagar, S.R. (1971) : "Rural Water Supply in India", *The Journal of Inst. of Engg. (India),* Vol.53, PH1, pp. 92-94.

Kuiper, E. (1971) : *Water Resource Development,* Butterwoths, London, p. 5.

Lenihan, J. (1977) : "Environment and Man (vol. 5)", *The Marine Environment,* Blackie, Glasgow and London.

Lohr, E.W. and Love, S.K. (1954) : *The Industrial Utility of Public Water Supplies in U.S.,* pt. 1, U.S. Survey Water Supply Paper- 1299, p. 639.

Majumdar, S.C. (1943) : *Rivers of Bengal Delta,* Associate Book Agency, Patna, p. 112.

Malik, D. (2001) : *Drought In Rajasthan Current Status of Situation and Interventions,* Health Environment and Development Consortium. p 5.

Mandal, R. B. (1976) : *Nature of Population Geography, Dimension in Geography.* Associated Book Agency, Patna, p. 96.

Marchall, C.L. (1972) : "Some Exercises in Social Ecology : Health, Disease and Modernization in the Ryukyu Island", *In The Careless Technology,* ed. M.T. Farvar and J.P. Milton, Now York, Natural History Press, pp. 5-18.

Ministry of Health, Government of India (1962) : *Public Health Engineering Manual and Code of Practices,* Section 1-As *Manual on Water Supply,* Government of India, New Delhi.

Minn (1995) : *Ground Water and Wells,* Johnson Division, UOP. Inc., Saint Paul, USA.

Misar, B.P. *et. al.* (1962) : Translation of "*History of Rajasthan by Col. James Tod*", reprint in 1987, publisher unique trades, Jaipur. p. 600.

Moore et al. (1966) : "Diarrheal Disease Studies in Costa Riça : 1, Plan and Method of Investigation", *Am. Journal. of Pub. Health,* 56, pp. 442-451.

Morgan, M.A. (1969) : "Overland Flow and Man", *Introduction to Geographical Hydrology,* Chorley, R.J. ed., Methuren & Co. Ltd, London, p. 56.

Mukeherji, B.K. *et. al.* (1962) : *Review of Work Done on Water Requirement of Crops in India,* Indian Council of Agriculture Research, New Delhi, p. 31.

Mullick, U.C. (1979) : *Development of Small Town for Planned Urbanization and Economic Growth : Place of Small Towns in India.* Inter-India Publication, Delhi, p. 55.

Murthy, R.S. *et. al.* (1979) : "Land Resources Inventory of Rajasthan by Using Colour Composites of Space Imagery", *Science Review,* CAZRI, Jodhpur.

Newspaper (2001) : *Dainik Bhasker,* dated 26 Feb. 2001.

Newspaper (2001) : *Hindustan Times,* dated 19 April, 2001.

Newspaper (2001) : *Hindustan Times,* dated 30 April, 2001.

Newspaper (2001) : *Hindustan Times,* dated 6 March, 2001.

NRSA (1980-82) : *Mapping of Wastelands in India from Satellite Imagery, Summary Report,* Rajasthan, Hyderabad.

Pesod, M.B. (1977) : "Surface Water Quality Criteria for Tropical Development Countries", In : *Water Wastes and Health in Hot Climate,* Feachem, R. *et. al.* eds., John Wiley, London, p. 56.

PHED (1988) : *Status of Drinking Water Supply in Rajasthan,* Water Mission Report, PHED, Jaipur, Rajasthan, p. 27.

PHED (1997) : *An Overview of Water Supply & Sanitation,* Public Health Engineering Department, Govt. of Rajasthan, p. 27.

PHED (1999) : *Annual Report,* Public Health Engineering Department, Govt. of Rajasthan.

Prasad, L. (1992) : "*Operation and Maintenance of Rural Water Supply in Bihar*", Conference papers on Rural Water Supply and Sanitation for Development Countries, Indian Water Works Association (R&D), Nagpur.

Prickett (1965) : "FAO Meeting Report", Food and Agriculture Organisation of the United Nations World Health Organization, 1965/10/1, p. 27

Rajvanshy, P.S. *et. al.* (1987) : *Study and Status of Environmental Consciousness in the Planned City Jaipur.* Rajasthan Pollution Prevention and Control Board, Jaipur. p. 1.

Ramaswamy, S. M. & Bakliwal, P. C. (1983) : "Ground Water Targeting of Banded Gneisses Complex, Rajasthan Through Remote Sensing, Proc. Nat. Symp. on Remote Sensing in Development and Management of Water Resources.

Rathore, M.S. (2002) : "Drought Management in Rajasthan: Some Long

Term Policy Issue", *Drought issue in Rajasthan,* Rajasthan State Public Administration Institute, Jaipur.

Romero, J.C. (1970) : *Ground Water,* New Orleans, Vol. 8, No.2, pp. 37-48.

Sahai, B. *et. al.* (1980) : "Remote Sensing of Water Resources in Panchmahal district," in *Contribution of Space Observation to Water Resources Management,* Pergamon Press, Oxford, p. 49.

Saini, S.K. (1996) : "Urban - Rural Water Supply in India", *Water Resources and Their Utilisation,* Vigyan Prakashan, Jodhpur, pp. 147-170.

Sally, H.L. (1968) : *Irrigation Planning for Intensive Cultivation,* Asia pub. House, Bombay, p. 69-94.

Sharma, S.K. *et. al.* (2000) : "Rain Water Harvesting For Ground Water Recharge In Maharashtra, *Ground Water Management Strategies In Arid and Semi Arid Regions,* National Seminar, Ground Water Dept. Govt. of Rajasthan, pp. 21-34.

Siddiqui (1970) : *General Report,* Geological Survey of India. p. 76.

Singh, J.B. (1976) : *An Agricultural Geography of Haryana,* Vishal Publication, Kurushetra, India, p. 378.

Singh, R.B. (1975) : "Rajput Clan-Settlement in Varanasi District", *National Geography Society India, Research Publication* No. 12, p. 15.

Singh, R.L. (1962) : "Meaning Objective and Scope of Settlement Geography", *National Geography Journal,* Vol. VII, p. 12.

Singh, S. (I976) : "A Note on the Photographic Technique for Location Ground Water Resources in Rajasthan Desert, India," *Geographical Journal of India,* Vol. XVII, p. 107.

Sinha, U.P. (1985) : *Planned Development of Resources in a Developing Region,* Inter India Publication, New Delhi, p. 38

Somani, L. *et. al.* (1983) : *Dictionary of Agriculture,* Agricole Publishing Agency, New Delhi. p. 292.

Stringer, E. et.al (1966) : *Geography of Resources,* Canvassel & Co., London, p. 25.

Surbrahmanyam, V.P. (1975) : "Water Balance Technique in Drought Climateology", *Vayu Mandal,* Vol.5, No.4, p. 75.

Trvedi, H.K. *et. al.* (1986) : *Impact of Green Revolution on Indian Farmers,* Deep Publications, New Delhi.

U.N. Water Resource (1976) : *Demand for Water, Procedures and Methodologies for Projecting Demand in the Context of Regional and National Planning,* ST/ESA/38.

UNDP/IBRD (1987) : *Community Water Supply,* The Handpump Option".

United Nations (1977) : *Vancouver Habitat 1977 Conference, International Drinking Water and Sanitation decade programme,* UN Resolution regarding Safe Water by 2000 AD etc.

UNO (1958) : *Water for Industrial Use,* UNO, New York.

Van Te Chow (1964) : *Hand Book of Applied Hydrology.* Mc Graw Hill Book Company, New York, p. 617.

Ward, R. C. (1978) : *Ground Water,* Chapman and Hall Company, New York, p. 19.

Washing, P H. (1949) : "Significance of Nitrate in Water Supplies", *Am. Water Work Assoc. J1.,* Vol. 41, no. 2, pp. 147-150.

Watt, J.et al. (1953) : "Diarrhoeal Disease in Fresno County, California", *Am. Journal of Pub. Health,* 43, pp. 728-741.

Web site : *http://www.investrajasthan.com/business/bis52828.htm*

Website : *http://www.cgiar.org/iwmi/home/rainwater.htm,* CGIAR online, Consultative Group on International Agricultural Research.

Website : *http://www.fluoridealert.org/news/india/rajasthan.htm,* web site of Fluoride Action Networks.

Website : *http://www.org.in/undmt/states/rajas,* United Nations Disaster Management Team.

Website : *http://www.rajamb.com/Geology.htm,* Rajasthan State Agriculture Marketing Board, Jaipur.

Website : *http://www.rajamb.com/soils.htm,* Rajasthan State Agriculture Marketing Board, Jaipur.

Website : *http://www.rajgovt.org,* Official Web Portal of Govt. of Rajasthan.

Website : *http://www.rajgovt.org/relief2002/,* Official web Portal of Govt. of Rajasthan.

Website : *http://www.rajirrigation.com,* Irrigation Department, Govt. of Rajasthan.

Website : *www.wsp.org/pdfs,* Jalvaani, News letters on Rural Water Sanitation in India.

White, G.W. (1977) : Pattern of Domestic Water Use in Low Income Communities, Water, Waste and Health in Hot Climate, in Ed. R. Feachem, John Wiley & Sons, p. 110.

WHO (1971) : *International Standards for Drinking Water*, 3rd. edition, Geneva, WHO.

WHO (1979) : Human Viruses in Water, *Waste Water and Soil*, WHO Technical Report Series No. 639.

WHO (1986) : "Guidelines for Planning Community Participation Activities in Water Supply and Sanitation Projects".

Wittman et al. (1967) : "An Evaluation of the Relationship Between Nutritional Status and Infections by Means of a Field Study", *South African-Medical Journal*, 41, pp. 664-682.

World Bank (1980) *Water Supply and Waste Disposal*, (Poverty and Basic Need Series), Washington DC.

Zdres, J. (2001) : *Hunger Admits Plenty,* A Few Facts & Figures (Oct to Dec 2000)

❑❑❑

Index

❑❑❑

Of Similar Interest

Water : Resources, Management and Economic Development

Pallavini Vashishtha

The book focuses on the interrelationship of water resources, management and economic development. It minutely analyses the status of water availability, accessibility and its uses. Further, it elaborates on various benefits like economic, health and several others, generated through water. Moreover, diverse indices like Water Poverty Index (WPI), Hydrological Water Stress Indicator (HWSI), and Social Water Scarcity Index (SWSI) have been calculated to assess the severity of water problem and also to ascertain their important determinants of water poverty and water accessibility.

This study also reflects over the water resources and economic development of Rajasthan with special reference to Jaipur district and it draws attention to the decline of water availability in the district to the extent that it is on the brink of turning into a water scarcity zone. It is an excellent guide to interdisciplinary and wide-ranging knowledge on water. It carefully draws comparison on water problems in background of India and other countries.

This book is an indispensable reference tool for all scholars and students of development economics, natural resources management, infrastructure planning, geography, demography and environment education.